Augustus Noll

How to Wire Buildings, a Manual of the Art of Interior Wiring

With many illustrations

Augustus Noll

How to Wire Buildings, a Manual of the Art of Interior Wiring
With many illustrations

ISBN/EAN: 9783743334885

Manufactured in Europe, USA, Canada, Australia, Japa

Cover: Foto ©berggeist007 / pixelio.de

Manufactured and distributed by brebook publishing software (www.brebook.com)

Augustus Noll

How to Wire Buildings, a Manual of the Art of Interior Wiring

HOW TO WIRE BUILDINGS:

A MANUAL

OF THE ART OF INTERIOR WIRING.

BY

AUGUSTUS NOLL, E.E.,

Member Am. Inst. of Elec. Engineers.

WITH MANY ILLUSTRATIONS.

SECOND EDITION.

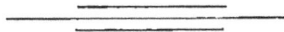

NEW YORK:
C. C. SHELLEY, PUBLISHER,
10 & 12 COLLEGE PLACE.

1894.

PREFACE.

It would be impossible to exaggerate the importance of good wiring as an element in the prosperity and permanence of the various electrical industries; and the writer of the present modest volume believes that his efforts to explain the true principles of good wiring and to point out the best methods will be welcomed cordially by electrical engineers, contractors, wiremen and the public generally.

The idea followed in inditing these pages has been to draw upon personal experiences ranging over a great many years, and to set forth in plain, simple language the things that must be done and the things that must be avoided. Every part of the subject is treated in such a manner that beginners, wiremen and others interested in the practical branches of the art of interior wiring can, with a little attention, understand readily why certain practices are preferable to others, and how a piece of work can be rendered safe and satisfactory. It is believed that practical advice from one who has himself tested by actual work and observation the value of every rule and suggestion here offered, will be very helpful. While much may seem novel, no statement is made that practice does not confirm, and no method is recommended that experience and common sense do not approve.

There is as far as possible all avoidance of abstruse technicalities in the descriptions, and the divisions have been made with a view to an easier comprehension of the different parts of the subject. A special series of drawings has been made for the book which will be appreciated, the author thinks, in several quarters ; and many of which have long been needed.

It is hoped that a perusal of this book may lead many workmen forward to the study of other books that they have hitherto found too technical, so that they will then be able to see how close is the relation between good work and sound theory. All that the author asks credit for is an earnest desire, of which this little book is the expression, for the perfection of an art to whose improvement he has devoted all his time and thought.

AUGUSTUS NOLL.

New York City, June, 1893.

CONTENTS.

HOW TO WIRE BUILDINGS.

CHAPTER I.

INTRODUCTION.

1. In the science of electricity the question of "wiring" is hardly considered, but in the application of electricity from a practical and commercial standpoint it is one of the most important and difficult factors. It embraces nearly every branch of what is termed "construction work." The student in electricity may have sufficiently mastered the theory of electrical currents, together with a knowledge of the different definitions and terms, to an extent that will enable him to "lay-out" wiring, on paper, in the form of a plan, showing the different circuits, sizes of wires, etc.; and while his work may be all that is desired in the shape of a plan, it will very often, even while the wires are being installed, fail of its purpose, owing to the conditions which exist in the building and which were not considered or known at the time of making the plan. The mere electrical student cannot thor-

oughly understand the "art of wiring," because its essential features are purely practical, and can only be acquired by experience and strict observation.

2. On the other hand, the practical wireman or beginner will make more rapid advancement if he will confine his studies to a general knowledge of electricity, sufficient to enable him to trace the direction of the current in dynamos, motors and appurtenances, and also in the wires, rather than fill his mind with all the different theories regarding magnetism and electricity, which will only tend to confuse him and make the different methods and systems of wiring appear more difficult and complex. When once the practical branch of wiring is sufficiently mastered, then will he be enabled to understand theory more easily, and he can gradually acquire a knowledge of the various terms, and of their relation to each other.

3. I have in these pages endeavored to explain, and illustrate, the more important features continually encountered in practical wiring, in as simple a manner as possible, eliminating all technicalities, so that this book may prove of interest, not only to the student, but also to the workman, in the "art of wiring" as practically applied in electric lighting and kindred purposes.

CHAPTER II.

General Considerations.

4. Electrical "construction work" covers such a vast field, that, not to make the work too lengthy, nor to exceed our proper limits, we will only treat of that portion which relates to the "wiring" of buildings, and such other branches incidental thereto, for electric lighting. It is, in this branch of the work, that the conditions are not fixed, and that those having charge of the work must depend largely on their own knowledge, judgment and ingenuity. There may be, for example, a change in the location or number of lamps; change in location of dynamo (in the case of an isolated plant), and numerous other changes which are liable to occur, and for which no provision had been made, when the work was originally "laid out."

5. Ordinarily, the foreman in charge should be competent: To draw a plan of each floor to be lighted; to understand and familiarize himself with the physical and other conditions, and from this data, proceed (leaving nothing to chance) to complete the plan of the wiring work and locate

the cut-outs and switches; locate the wires, and
decide on the method of distribution, and locate
the feeding points between the Feeders, Mains,
and Lamp Circuits; to decide, on the most satisfactory percentage of loss of energy to be allowed
in the different portions of the wiring system, so
that the Difference of Potential, or pressure at the
lamps, will practically be equal throughout the
building.

6. The foreman must also be able to compute,
according to the per cent. of loss, the cross-section
or size of wire necessary for use in the different
parts of the wiring system; to determine the most
satisfactory kind and length of fixture, and, when
the wiring work is completed; to set up and run
(in the case of an isolated plant) the dynamo, and
connect same, with the wiring and the instruments
usually provided; to compute the width of belt
necessary and also the diameter and face of driving
pulley if belted to a pulley on shafting, and in
case of necessity run an engine; and, finally, to
locate and repair all defects in the wiring, and
also slight defects in the dynamo, etc.

7. These items form the necessary qualifications
to ensure satisfactory and lasting results. The successful operation of the plant depends, to the greatest extent, on the manner in which the wiring was

"laid out," and electric lighting companies can trace the greatest part of the dissatisfaction on the part of their customers to the defective manner in which the wiring work was installed.

CHAPTER III.

LOCATION OF CONDUCTORS.

8. It is sometimes the case in installing wires in buildings that, although the best grade of materials, of the several kinds usually employed for the work, was used, the results have been unsatisfactory; the workmanship and appearance may have been all that could be desired, and yet in a short time after the current has been in use the work breaks down, and short circuits, grounds, leaks, etc., are of common occurrence. The fault is usually due to the location of the wires.

9. The following adverse conditions should be avoided as much as possible, and where it is impossible to do so, proper safeguards, according to the condition, must be provided: Excessive moisture, atmospheric changes, extreme variations of temperature, extremely high· temperature, gases, acids, lye, lime, cement, etc., and last, but one of the most important conditions, mechanical interference.

10. In dealing with the first class of conditions the best plan, of course, is to shun dangerous

places, but the location of the lamps may make it impossible. Generally, the best grade of materials and superior workmanship will lessen the chances of trouble, and with the high grade materials, as now manufactured, very little trouble should be experienced.

11. By the use of conduits (see chapter on conduits) and high grade moisture-proof wire, providing a separate conduit for each wire, the results have been most satisfactory. Special care should be taken to have the joints, on both the wire and conduits, water-tight, and equally as well insulated as the remainder of the wire and tube forming the circuit. The cut-outs and switches should be grouped and located in places free from the deleterious conditions above noted.

12. The lamps, sockets and fixtures should be designed with a view to protecting the electrical connections contained therein. Keyless sockets should be used, and the lamps should be controlled by switches. The lamp should be provided with an extra safeguard, arranged to exclude moisture and gas. The design of these is generally similar to that of a fruit jar, or bulb of thick glass hermetically sealed by means of rubber bushings and metallic cover. If the style of work is what is

known as "open" or "exposed" work, the use of conduits, in many cases, will be unnecessary, but care must be taken to keep the conductors free from contact with the building, which can usually be accomplished by the use of suitable insulators.

13. In dealing with conditions that come under the head of mechanical interference, care must be taken to locate the conductors and all appurtenances used in the wiring work in such places that while they are of easy access to persons in charge, they are still inaccessible to inquisitive, uninformed and malicious persons.

14. When installing wires in a building in the course of erection, be careful to isolate and prevent contact with the work of other mechanics as much as possible. As the building is in a rough state, the wires must be handled with care to prevent abrasions in the insulation. The nature of the material forming the insulation will not admit of rough treatment, such as dragging along the rough floors, or across piles of brick and mortar, etc. In short, the wires should be so treated that the insulation covering them will not be bruised, cut or perforated in the slightest degree. In an otherwise perfect insulation the most minute disarrangement of its parts will often impair its use-

fulness and permanence, and while, in some cases, the defect will at once become evident, still, in other instances, the fault may not appear until after the work has been in use for some time, when remedying the defect will cause expense and extreme annoyance, and will necessitate the cutting of the plaster on the side walls or ceilings.

15. The destruction of the insulation, due either to natural deterioration, or other causes, is the greatest source of trouble in electric lighting (see chapter on electrolysis). Hence the use of inferior insulated wires is always false economy.

16. Keep the work clear of and at some distance from places where exposed wood-work is to be fastened, such as the trim of doors, windows, chair-rails, picture mouldings, base-boards, etc., as these places will be plastered, and should the wires or conduits be located at these particular points, their liability to being cut by nails is very probable.

17. When wires are placed under the floors or between partitions, they should be separated from each other and fastened beyond the reach of the floor nails or lath nails. Inaccessibility of the conductors should be avoided as much as possible, and concealment of same should only be resorted to where definite channels, such as conduits, have

been provided. Do not sacrifice necessary safeguards for a neat appearance, and do not "fish" electric light conductors, in the same manner as burglar alarm, annunciator, electric bell and such like conductors usually are. In the latter cases the wires are "fished" or threaded here, there and everywhere, so that they may be hidden from sight, but in the case of lighting circuits, no conductors should be inserted in a channel unless the exact condition of same is known. In this case you are dealing with *horse-power*, and nothing must be left to chance. Wires installed in buildings must not be treated according to the apparent conditions. In every case a personal inspection by one experienced in this branch of the work is the only safe method.

18. Marble and tiled floors, etc., should be avoided as a place to locate wires whether conduits are used or not. In cleaning such floors, sulphuric acid is made use of; and materials of this nature absorb moisture, which will, in addition to the action of the acid, create another source of trouble. Regarding moisture and like conditions, no building for electric lighting purposes can be assumed to be dry. While it may have been dry at the time of installing the wires, the probability of this condition being changed is so great that as

a safe rule it should be treated as damp and the corresponding grade of material used as though it were continually damp or wet, because the water-pipes may leak or burst, or the bath-tub or basins may overflow.

19. When wires are encased in wood mouldings, the latter should consist of a grooved back board and a tight fitting cover, and in stores and similar places the ceilings should be kept clear of same as much as possible. A separate groove should be provided for each wire, and in no case should wires of the same or different polarity, forming different circuits, be placed in the same groove. That style of wiring should never be made use of in places where adverse conditions exist.

20. In hotels, office buildings, residences, etc., locate all the appliances, main and feeding lines, in hallways or closets, so as not to have to enter rooms. The annoyance of interference, both to the inmates and workmen, may thus be obviated while the work is being installed. It will also obviate annoyance in the future, when repairs, renewals or alterations may be necessary. In workshops, fac-tories, etc., the wires are usually held to the ceil-ing by means of cleats, or fastened to porcelain knobs. The wires should always be located in

such places on the ceiling as are clear of shafting, belts, etc., and where the work can proceed without interruption either to the wiremen or the hands and machinery employed on the premises.

CHAPTER IV.

Division of Circuits and Distribution of Current.

21. When the building has been inspected and the various conditions in the different parts have been noted, the question of distributing the current then demands attention. The most important considerations are the equalization of pressure throughout the different lamp circuits, and to divide the circuits according to the manner in which the lamps will be used. In a building where the wires are to be concealed (new structure) the best mode of procedure is to locate the various closets or boxes for the reception of the cut-outs and switches. The location is governed by the number of lights, and size and shape of the building. Generally from four to eight lamp circuits are carried to each box. The box should be, as nearly as possible, in the centre of the space, the lamps in which are supplied with current by the circuits centering at this box. After locating the box and dividing the lights into circuits, the next step is to decide upon the method of dividing and connecting

the main and feeder wires. The number of mains and feeders depends upon the size of the building and number of lights. Locating the connecting points between the mains and feeders depends upon the location and number of lights at each box or ramifying point, the distance between the various feeding points and the use of the lights. That is to say, in one part of the structure the lamps may be used continuously, and in another part only a few out of a great many may be used at any one time. The building should be cut up or divided into sections. Either each floor or each side of the building should comprise a separate circuit, according to the size of the building (see chapter on explanation of distribution). If divided perpendicularly, one or more mains feeders are provided, extending from the main switchboard in the dynamo-room to the central point between the feeding or connecting points on the line of the mains. The mains are usually run from the top to the lowest floor, and at each floor connections are made to wires which carry the current to the different lamp circuits. These floor mains connect to one or two cut-out boxes. If they are located somewhat closely together, then only one set of main wires are required; but should the distance between them be great, the floor main must extend to a point midway between them, and again be divided, so that a branch from each will connect with the lamp circuit independent of the other.

Should the structure be composed of numerous floors, it will be the best plan, instead of connecting the main feeder from the dynamo-room to the mains in the central point, to provide a set of subfeeders starting and connecting with the main feeders at its terminals, which should extend both ways and connect with the vertical mains at a point about one-quarter of the whole distance from the end. The idea of feeding in this manner is to equalize the lighting; that is, to balance the electrical pressure and the number of lamps in use, so that the variations constantly occurring will not affect the pressure to an extent where the life of the lamps will be shortened.

22. The various systems of feeding all tend to this, as is shown in the plans (see chapter on same). From these it will be seen that the circuits are constantly being divided into smaller sections, so that the electrical pressure is practically the same throughout, and that each branch is, to a certain extent (electrically), independent of the others.

When "laying out" the feeding lines, not only should the work be done with a view to equalizing the pressure on the wire with the full load on, but also for periods when only a few lamps are in use on any one branch, while in still another all the lamps were in use, so that the variation of pressure on the conductors connecting with the few lights will be 'reduced to a minimum. To accomplish this, it is at times necessary to install two or more sets of feeders, connected at the same main conductors, but at different locations. Again, in other instances, such as a large isolated plant, or where the lighting covers a large area, it is necessary, in order to equalize the electrical pressure at the lamps, to provide and connect in the main feeder circuits regulating devices for controlling the pressure.

23. In a plant where the loss of electrical energy in the conductors has been computed at a high rate, the pressure at the lamp will be at the rated amount only when all the lamps are in use; and, should only one-half be in use on any one section, the percentage of loss in that section will be only one-half of that in which all the lamps are in use. Consequently, the pressure on the wires in the section where only one-half are used will be greater than that at which it was rated, and the result would be excessive breakage of lamps.

To overcome this defect is the function of the pressure equalizer. It may not be amiss to here explain what work the *equalizer* performs and the difference between it and a pressure *regulator*.

24. A PRESSURE REGULATOR generally consists of a coil or numerous coils connected in the field or magnetic circuit and is usually (at the present time) controlled automatically, although regulating instruments are constructed which can be governed by hand. The coils are placed directly in the magnets, as in a compound wound machine, or by the use of magnets, all or any part of the coils can be thrown in or out of the field circuit. As the lamps are connected or disconnected on the various circuits, so will the strength of the pressure increase or decrease. As the pressure increases, the coil or coils are thrown in circuit, and, according to their resistance, the pressure is decreased until it is at its normal strength. The pressure is kept at a constant strength at the dynamo brushes.

Should there be two or more independent sets of feeders carrying current to different parts of the structure, and on one set all the lamps be in use, the pressure at the lamps in this section would be normal; but if on another only a few out of a large number of lamps were in use, the pressure would be greater than that at which it was rated. The pressure required for the lamps, plus the percentage of loss in the conductors (assuming no loss

on the line with only a few lamps in use), would
equal the pressure at these particular lamps, and if
the percentage of loss was computed at 10 per
cent., then this amount would represent the increased pressure. To overcome this and equalize
the pressure on this particular line without affecting the pressure on the other circuits, the EQUAL-
IZER is used. It consists of a coil or numerous
coils, divided in sections and connected to plates
fastened to a slab of slate known as the equalizer
face board. The coils are thrown in or out of the
feeder circuit by means of a movable connecting
plate fastened to a handle. In the case of a pressure regulator, the diameter of the wires is small,
as they only carry the amount of current carried
by the magnet wires, but the equalizer being connected directly in the circuit, it must have sufficient
carrying capacity to carry the full amount of the
current necessary, according to the number of
lamps on the circuit, without heating. The size
and length of the wire forming the coils depend
upon the number of lamps and the percentage of
loss allowed in the conductor. Where several sets
of feeders are used, each set should be provided
with an equalizer. The equalizer is connected on
one side of the circuit only, in a similar manner to
a one-pole switch. Throwing the coils in or out of

the circuit is governed by the indications, as shown on the scale of the pressure indicators, which are provided and set in a place near the equalizer face board. The indicator is connected in the circuit by a set of pressure wires which connect with the circuit at the center of distribution, and as the pressure varies at that point, the amount of variation will be indicated by the needle as it travels across the scale of the indicator, so that should the instrument indicate excessive pressure at the center of distribution, the current can be brought to normal strength by connecting or throwing in the circuit sufficient coils, the combined resistance of which will decrease the pressure to an extent where it will be equal to that required by the lamps to keep them up to their rated candle-power.

25. To return to the subject of distribution again. The number of circuits should be as many as practicable; the greater the number, the less the variation that can ensue.

The system, as explained, is known as the "panel" or "grouping" system, and is principally used in new structures, office buildings, hotels and such structures. In factories, stores, etc., the method of distributing the current in the main and feeders is the same, but the lamp circuit wires, instead of being grouped, are connected at the nearest point with a ceiling or floor main which extends through the whole length of the building. The lamp circuits terminate in a cut-out which is connected to the mains. The wires are either exposed or covered by grooved wood mouldings. In all structures where the conditions are such that all the lamps on certain circuits are used at a given time, it is preferable to connect them to a special, independent feeder, so that the pressure on the circuits connected to the lamps used at intervals can be more nicely regulated.

The lamps that are usually turned on and off at stated times, or which are only used on special occasions, are generally the entrances, hall, stairways, reading and waiting rooms, etc.

26. In all wiring installations the public lights should form a separate circuit.

The public lights are generally those in public parts of the building, and also generally include the engine-room, cellar, etc.

All residences, hospitals, etc., should be provided with a night circuit, having one or more lights in the hallway on each floor, and connected in such a manner that they can be connected or disconnected from one or more places, and can also be connected to the burglar alarm and light auto-

matically in a time of danger. This circuit is a
great accommodation in case of sickness, sudden
alarm, etc.

CHAPTER V.

Loss of Electrical Energy in Conductors.

27. The expression "percentage of loss," when used in connection with electric lighting, signifies the amount of electrical energy expended in the conductors and is due to their resistance. The rate is governed by the local conditions, but, usually, wiring connected to mains from central stations is figured on a basis of two per cent. loss, and wiring for isolated plants at five per cent. loss. To a certain extent, it is governed by financial conditions, that is, the percentage of loss is increased in cases where fuel is cheap or when water-power is used. As a general rule for ordinary isolated plants it may be stated thus: Decrease the loss and the cost of maintenance is lessened, but decrease the percentage of loss and the first cost is increased. It is preferable to slightly increase the first cost by wiring at a low rate of loss because the decreased cost of maintenance will more than offset the difference in first cost in generating the current by the saving in fuel. The liability to exces-

sive lamp breakage is also obviated when the percentage of loss in the conductor is low.

28. The first cost, assuming the boiler, engine, etc., to be already installed, consists of the generating apparatus and its appurtenances, wire, appliances, fixtures, lamps and cost of labor for installing same. With proper care, the generating apparatus, its appurtenances, wire appliances and fixtures, will not depreciate to any considerable extent. Therefore, the cost of maintenance consists chiefly in the amount of fuel necessary for generating current and the renewal of lamps. The first item is governed by the amount of current required to maintain the rated candle-power in each lamp. (See formula electrical energy.)

The renewal of lamps depends largely on the breakage due to the action of excessive current. When the pressure is normal, the action of the current in passing through the carbon is such that the carbon is gradually decomposed and disrupted to an extent that finally the carbon breaks. Under ordinary normal conditions the lamp will give continual service for from 600 to 1,500 hours. This is termed "the life of the lamp," but should the electrical pressure vary so that at certain times it is greater than the strength required, the carbon will be overtaxed; the strain will be too great, the disruptive action will be increased and the carbon will become defective and weak, and usually in a short time it will break. The amount of coal consumed is governed by the amount of electrical energy generated in the dynamo, and if ten per cent. of the electrical energy is required to overcome the resistance in the conductors, then practically ten per cent. of the total amount of fuel used is the cost of carrying the current from the dynamo to the lamps. From this it will easily be seen that the question of percentage of loss in the conductors must be most carefully considered, not only from an electrical but also from a commercial standpoint. Another important consideration is to have the percentage of loss distributed properly in the different conductors forming the circuit. The most general and satisfactory method is, assuming the loss to be five per cent., to allow one-half per cent. in the lamp circuit, one per cent. in the floor mains, one and one-half per cent. in the vertical mains and two per cent. in the main feeders.

While the term "percentage of loss" indicates a loss of energy between the dynamo and lamps, commercially it is considered as the cost of transmission, and, if intelligently considered, will not be treated as a waste of power, but as a legitimate item in the cost of lighting. No energy can be transferred without loss or cost.

29. To demonstrate the principle more clearly, assume a lamp located at a distance of 1,000 feet from the dynamo. The resistance of the lamp (the resistance of lamp is always figured at the time the lamp is hot, or at incandescence) is 90 ohms, and it requires, to bring same to its rated candle-power, a current of one-half ampere at 45 volts. We will decide to lose ten per cent. of the electrical energy in the transmission of the current from the dynamo to the lamp, in the conductors. If the difference of potential at the brushes of the dynamo = 45 volts, the difference of potential at the lamp would equal 45 volts — ten per cent. loss = 41.5 volts. But we wish to get a pressure of 45 *volts at the lamp*, which is necessary to bring the lamp to its rated candle-power. Therefore, 45 volts only represent 90 per cent. of the pressure. Thus 50 volts = the pressure at the dynamo brushes. This lamp having a resistance of 90 ohms; 2,000 feet of wire being required to connect same with the dynamo; ten per cent. being the loss expended in the conductors, and 90 ohms representing 90 per cent. of the total resistance of the circuit, the 2,000 feet of wire must equal, in resistance, ten per cent.

of the entire circuit, which is 10 ohms. Now, to
find the electrical energy expended in the conductors and lamp expressed in watts (see formula):

$$\frac{E=45}{R=90} = C = \tfrac{1}{2} \text{ and } C \times E = \text{watts} = \tfrac{1}{2} \times 45 = 22\tfrac{1}{2}$$

watts expended in lamp, and $\tfrac{1}{2} \times 5 = 2\tfrac{1}{2}$ watts

expended in conductor.   Or, $C^2 \times R = \text{watts.}$

Therefore,

$.5 \times .5 \times 90 = 22.5$ watts expended in lamp, and

$.5 \times .5 \times 10 = \;2.5$ "   "   " conductor.

In both of the preceding instances it was shown
that the lamp required one-half ampere at 45 volts;
consequently a greater pressure was required at
the dynamo to allow for the "drop" in the conductors. In this last case it required 5 volts,
making the difference in potential at the brushes
of the dynamo 50 volts, the combined resistance of
the lamp and conductors being 100 ohms. Consequently the current consumed on the entire cir-

cuit $= \dfrac{50 \text{ volts}}{100 \text{ ohms}} = \tfrac{1}{2}$ ampere. But this $\tfrac{1}{2}$ ampere

is developed at a pressure of 50 volts. Therefore,
it required more power in watts, as 50 volts $\times .5$

ampere = 25 watts = the power required for lamp and conductor.

From this it will be seen that the question of percentage of loss must be carefully considered, not only from a financial, but also from an electrical, standpoint.

CHAPTER VI.

Plans.

30. The plans generally used for showing electric light wiring work are what is known as Floor and Elevation plans.

The Floor plan represents everything as being flat, and on it are marked the outlets, showing the number of lamps to each, and whether from the side or the ceiling; the size and location of the wires forming the lamp circuits and the lamps which are to be connected to each; also the size, style and location of cut-outs and switches.

The Elevation plan should show the number of sets of feeders and mains, the exact location of points of connection between the feeders and mains, and between the mains and sub-mains which connect with the lamp circuits either at the grouping point or where the cut-outs are located along the ceiling, as in factories or in general open work. The style, size and location of main cut-outs and switches should also be shown. In the ordinary two-wire system it is usual to show only one wire, it being understood in this case that the wires forming the other side of the circuit will run parallel with the one shown on the plan, and that the cross-section is to be the same.

31. In formulating a plan of wiring it is best to be familiar with the nature of the structure, so that the wires may be fastened in places where the mechanical conditions are favorable.

While the "runs" should be as straight and short as possible, still the drilling of walls and partitions must be considered. In some cases it is cheaper to lengthen the circuit; that is, run the wires at an angle to those already in place, so as to form a turn and run somewhat out of the way, or so as to obviate the necessity of cutting through a thick brick or stone wall; or, in another instance, to have the wires free from the interference of steam pipes, leaks of pipes, etc.

It is preferable to make an examination of the premises before laying out the work on the plan.

The plan should be arranged to show all the different circuits and appliances as plainly as possible.

A good method is to designate all the different cut-outs and switches by a letter of the alphabet, and on a separate sheet or on the margin of the plan, refer to same. The quantity of each appliance, and the amount of each size of wire, will in this way be an easy matter to estimate.

32. When the work is completed, the plans, together with any alteration, either in the number or change in location of lights and change of location or size of wire, should be returned to the company. A report on the changes, and by whose authority the same were made, should also be submitted.

CHAPTER VII.

"Conduit Wiring."

33. When concealing wires, either by threading them between the partitions, or under the floors, or embedding them in the plaster, so that when covered they are totally inaccessible, the insulation is subject to deterioration and impairment due to various causes which may disarrange the lighting system of the light and cause failure or a fire.

It is well known that the best insulated wires will withstand the action of lime, cement, alkalies, etc., for only a short time. The insulation will then be useless for the purpose. The natural outcome is that the circuit must be rewired, which usually means but a choice of evils.

The plaster must either be cut, or the carpets, floor, etc., must be torn up to get at the wires. The wires can be run on the surface and covered with wood moulding, but this is not desirable in decorated places. These same conditions confront us when additions or alterations are necessary. Buildings wired for so-called "future use" prove, in the majority of cases where it is proposed to

use the wiring a few years after, to be unfit to turn
the current on, and the re-wiring generally costs
more than the original work. Tinkering with the
wires already installed would simply be false economy. The materials used in electric lighting are
affected by conditions which are not considered in
gas lighting, but there is no reason why the work
should not be equally as permanent and successful
as good gas piping. This can only be accomplished
by arranging all the conductors and appliances in
such a manner that the whole wiring installation
is accessible, but defacing the walls and ceiling is
objectionable and must not be attempted.

34. It is therefore necessary to locate the wires
in places out of sight yet all accessible. To do
this it is necessary to form channels in the walls
and under the floors for the reception of the wires
exclusively. The cheapest, safest and most prac
tical method is to provide conduits and boxes for
purposes of branching and as receptacles for the
cut-outs and switches. These boxes are also used
for the purpose of withdrawing or inserting wires,
and may be considered as a hand-hole, similar to
a manhole used for underground work.

The conduits not only form an extra mechanical
protection to the wires, but also exclude those
elements which generally render the insulation on

the wires worthless. Ordinarily it is impossible, except at a great expense, to change the system of lighting from the two to the three-wire system, or *vice versa*, or, from 100 to 50-volt lamps. The use of conduits admits of any practical change either in the system or in the number of lights.

Buildings can be equipped with conduits (instead of wiring) for future use, and the insertion of wires is unnecessary until the actual time of service, and at a very small increased cost.

For electric lighting purposes, no wire should be inserted in undefined channels, such as between floor and partition, because the conditions existing in the channel are unknown; they may change at any, time, and the result is a matter of chance and peril.

35. The conduit should be constructed of such materials as will meet the requirements of the purpose, and the diameter of the bore should be ample for the conductor. They should be installed in a manner somewhat similar to that of wire. The most essential features to be considered are : the completeness of system as to points of accessibility and absolute continuity of the tubes or wire-ducts.

The conduit system consists of joining together lengths of the tubes, using elbows where turns and bends are necessary. When the length or number of turns or bends in a circuit are such that the insertion of the wire becomes difficult, angle or "fishing" boxes must be provided so that it will be an easy matter not only to insert, but to withdraw the conductor. Junction boxes, according to the direction and number of branches, must also be provided. The boxes act as a receptacle for the cutouts, keeping them out of sight, and imparting a finished appearance to the work.

The conduit must be continuous ; from outlet to outlet, from box to outlet, and from box to box. Care must be taken to have all joints, either at the coupling or the box, water-tight, and well insulated. The inner surface of the tube must be in alignment throughout ; that is, no shoulders or impediments should be created, so that the insertion of the wires can be done quickly and without trouble. In a system of conduits embracing all the favorable features, the insertion of wires in the conduits can be deferred until the building is actually completed. The cut-outs and switches can be inserted and connected at the same time. The metal parts will remain bright and clean, and the troubles usually encountered in new buildings, due to corrosive connections, are in this case obviated

The insertion of drawing-in strings at the time of installing the tubes must not be relied upon. The

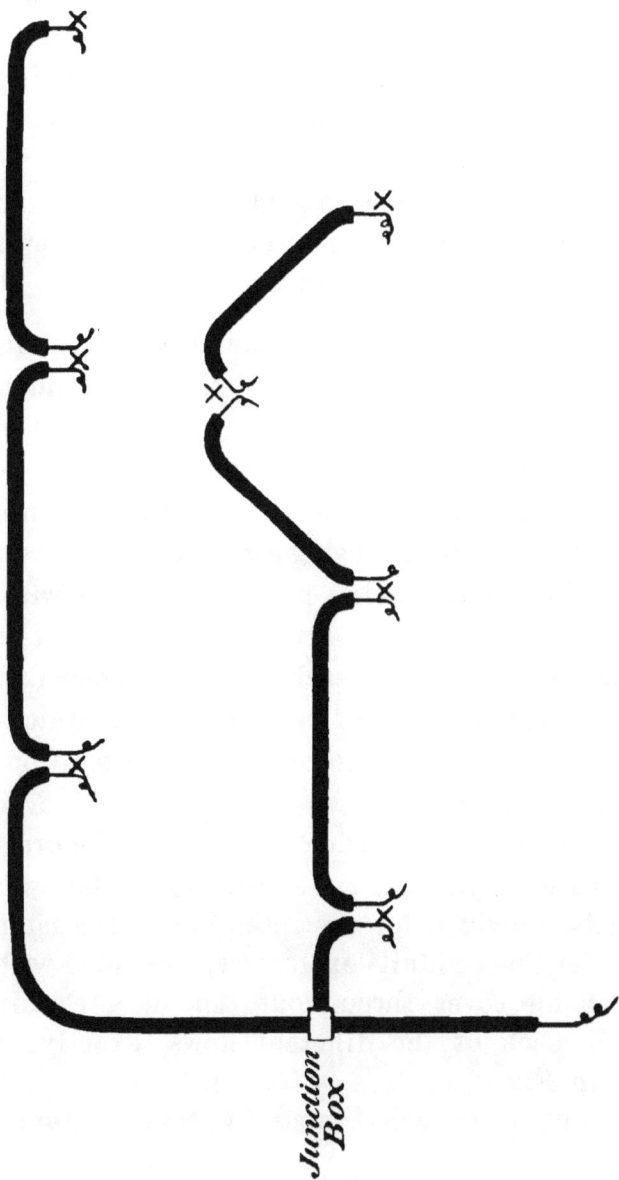

Junction Box

FIG. 1.—CONDUIT WORK.

tube must be considered empty, and installed in such a manner that should the drawing-in string break, the insertion of the wires can still be easily effected.

Where the conditions are adverse to ordinary wiring, the use of conduits carefully installed will generally meet all the difficulties.

36. The conduits and boxes should be located in the halls or public parts of the building as much as possible, so that additions and renewals can be made without annoyance to the tenants. The bends and turns should occur as near the ends of the lines of conduit as possible, it being easier to insert wires, more perfect control being had over the fish wire. The conduits should be so arranged that a point of entrance is obtained at each outlet, as shown in Fig. 1, so that the work of insertion may be quickly accomplished. It also admits of alterations being made quickly.

The system of distribution is similar to the ordinary methods of wiring. A recess in the wall should be provided for the reception of the main and feeder line conduits exclusively, provided with a detachable cover throughout; or at a certain point on each of the different flows, exactly as shown in Fig. 2.

In the figure showing tubes in recess A repre-

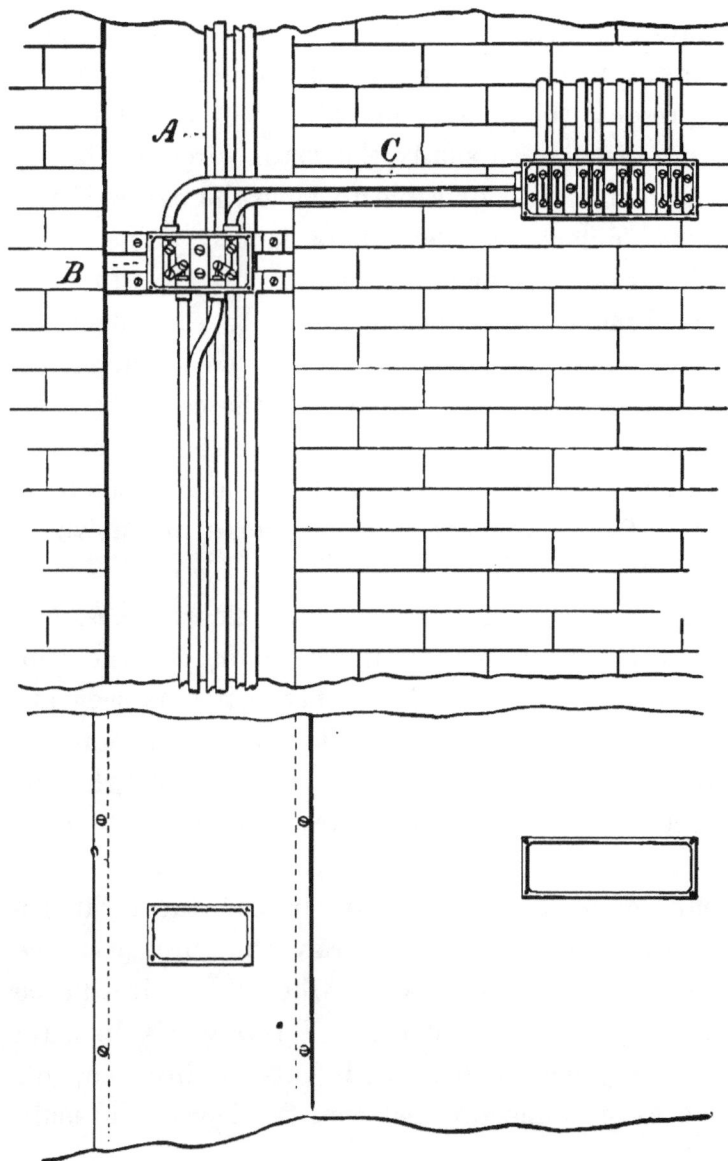

FIG. 2.—CONDUIT WIRING.

sents the main tube. *B* represents the main floor cut-out, inserted in a floor main junction box, *C* represents the floor mains which connect with the lamp circuits. Usually the main wires are large, which make it advisable to have the cover of the recess detachable the whole length. The main wires can be so arranged that a pair are run from the dynamo switchboard direct to each floor, or the system of distribution can be so arranged that the wires can be looped in the conduit, from floor to floor, thus relieving the conduit of the strain due to the weight of heavy conductors. The connections can be made as shown in Fig. 3.

In Fig. 3, *A* represents main floor boxes, one on each floor, to which are connected the conduits for the main wires. The wires as seen are cut at each floor, so that instead of starting at A^2 and drawing the wires through the conduit to A^3, it is only drawn from box to box, and the conductor is made continuous at the cut-out connection. Hence, if in the future it becomes necessary to withdraw the wire from the conduit, between any of the floors, it can be quickly and easily effected. *B* represents the main feeder junction box with cut-out inserted, for means of connection between the feeder and main

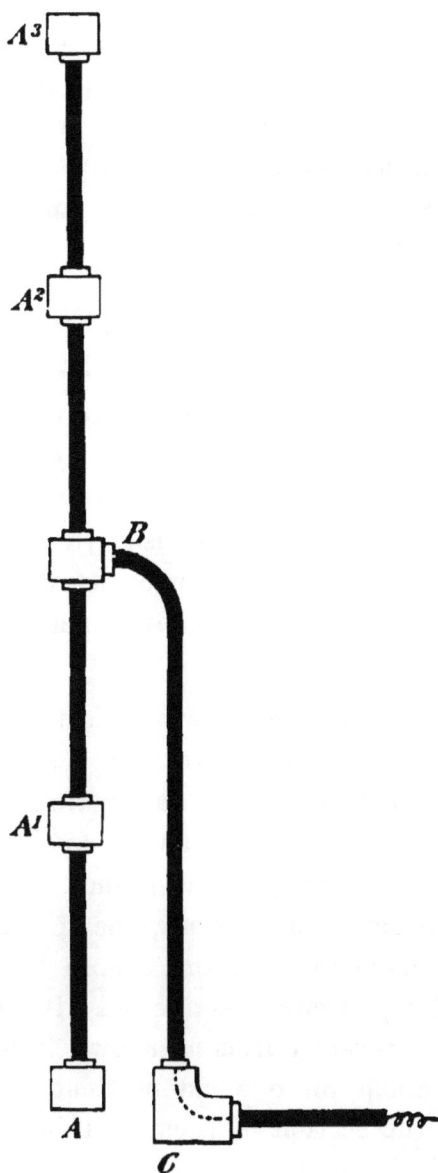

FIG. 3.—CONDUIT WORK.

wires. As this wire will generally have a large cross-section, it will be necessary to provide an angle-box at *C* (which is at the bend where the feeders assume a perpendicular position). The angle-box obviates the necessity of elbows, and at the same time performs the function of a "fishing box."

The lamp circuits are divided in a similar manner to that made use of in ordinary wiring. All the circuits are brought to a central or grouping point, and the conduits terminate at the box enclosing the cut-outs, as in the "Panel" or "Closet" method. Judgment must be exercised in dividing the circuits so that the run from outlet to outlet will not be too long or will not contain many turns or bends.

37. One of the most simple and satisfactory methods of floor wiring in conduiting, is shown in Fig. 4, which represents all the floor main lines of conduits located in the corridor, and at a point near the ceiling. *A* represents single branch junction boxes, fitted with a branch cut-out, one of each being located opposite to the room, whose lamps it controls. The lamp circuit is connected with it and is looped from outlet to outlet as shown in Fig 5, keeping each room on one independent circuit. The mains for the section of rooms will be looped

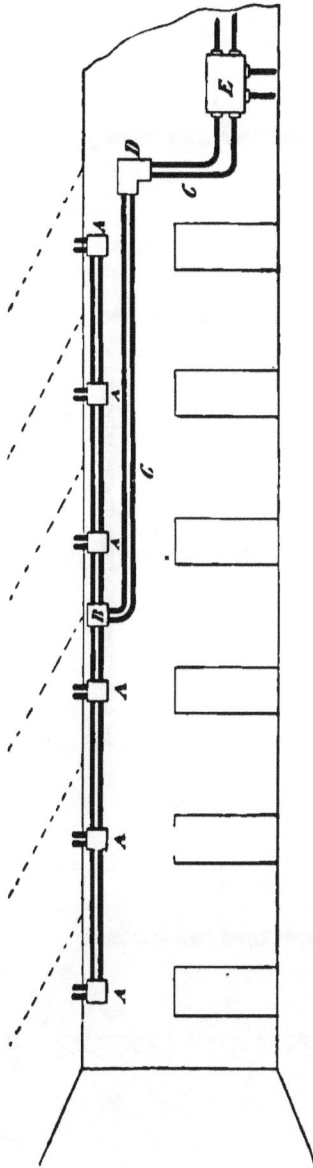

Fig. 4.—Conduit Work.

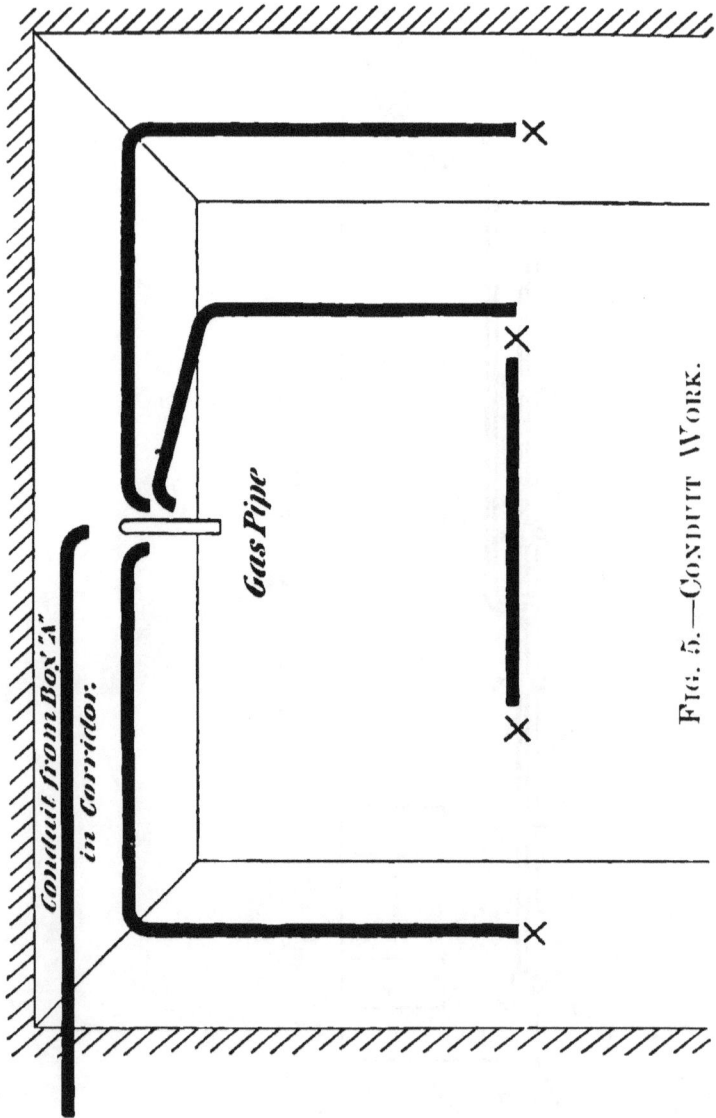

FIG. 5.—CONDUIT WORK.

in a similar manner from A to A, so that provision is made for easy manipulation. B represents the main cut-out and junction-box for the section through which the floor main C connects with the section main; D represents an angle-box, for use at the turn, and for making easy the insertion of the main wires in conduit C. E represents the main floor junction-box and cut-out.

By the use of this method, the pressure on the wires can be equalized just as nicely as in the "Panel" system. This method tends to shorten the distances between feeding points, manipulation is easier, and while the cost of material is not increased, the amount of labor is considerably decreased.

38. When the lighting is for general illumination purposes, that is, when the lamps are turned on or off at the socket, as in office buildings, etc., the lamp circuits can be installed in various manners, as shown in Figs. 6 and 7. The advantages obtained are, that should the wire in the cut-out (cut-out link) fuse, it would only disconnect a portion of lamps located in any one room. Also, it will sometimes save material and time, by shortening the length of the circuits, and avoiding turns and bends.

39. The joints must receive the same care as the

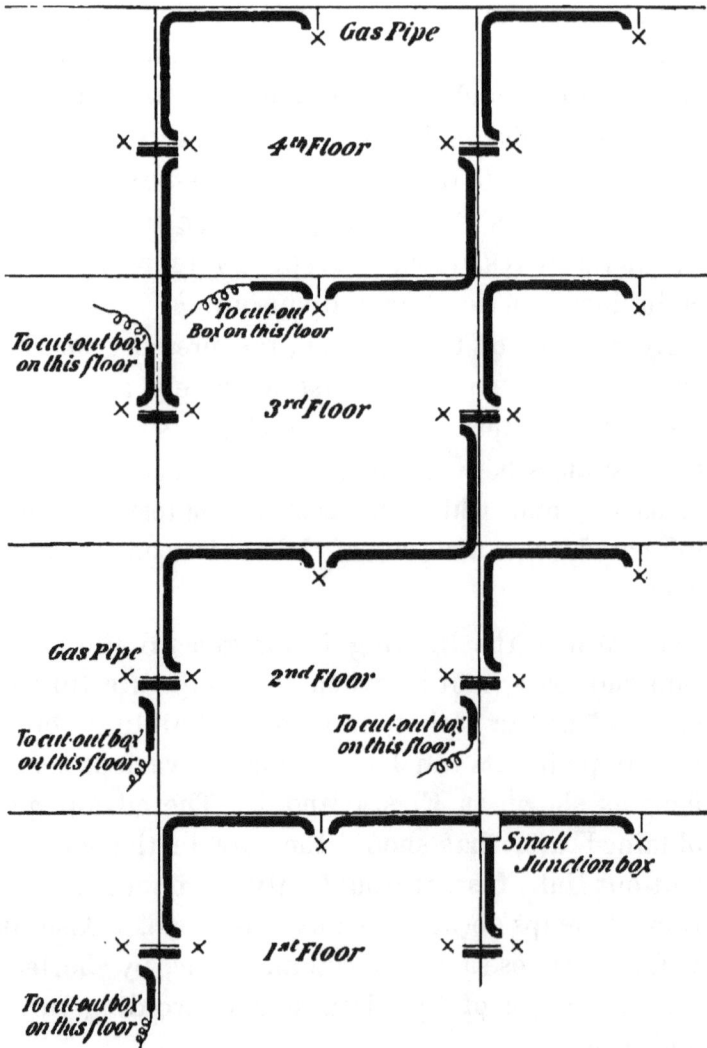

Gas Pipe

4th Floor

To cut-out
Box on this floor

To cut-out box
on this floor

3rd Floor

Gas Pipe

2nd Floor

To cut-out box
on this floor

To cut-out box
on this floor

Small
Junction box

1st Floor

To cut-out box
on this floor

FIG. 6.—CONDUIT WORK.

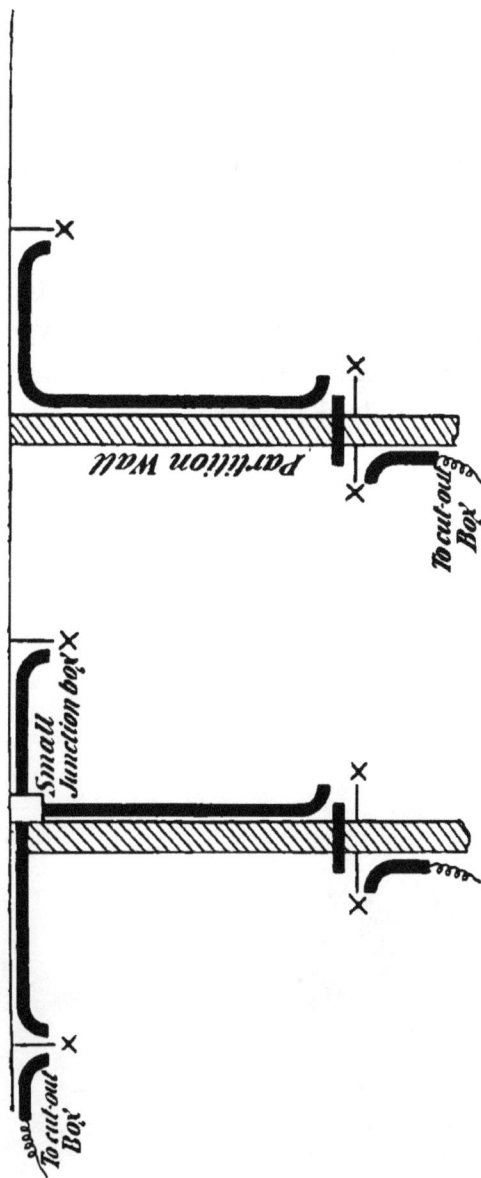

FIG. 7.—CONDUIT WORK.

joints on the wire, and the conduits should be handled in a careful manner. In no case must a hole or break in the conduit be patched.

Not only must the continuity of the conduit be maintained throughout, but care must be taken to maintain its circular form also.

In concluding this chapter it may be stated in short, that the use of conduits afford the only safe, reliable, and permanent method of wiring for electric lighting and kindred purposes.

CHAPTER VIII.

SWITCHBOARDS.

40. It is preferable to make a plan, to scale, showing all the appliances it is intended to place on the switchboard before constructing it, so that sufficient space and insulation may be provided between conductors of different, or even of the same polarity. The appliances and conductors should be so arranged that, whether for renewals or repairs, the different circuits can be disconnected without disturbing the remaining circuits, and without jarring or dislocating the board. The faceboard should be constructed of materials that are fire and moisture-proof, such as marble, slate, etc.

Sufficient space should be allowed between the back of switchboard and the wall of the room for purposes of inspection, repairs, etc., and also that should the wall be damp, the switchboard will be free from contact, and will not have water accumulate on it. Theatrical and converter switchboards should be provided with a covering either in the form of a door, or a roll, similar to that of a roller-top on a desk.

41. All switches and cut-outs should be equipped with a suitable name-plate, designating the particular circuit they control. The name-plate should either be fastened on, or directly under, the appliance. In all switchboards, especially those used in theatres, the switches, cut-outs and other appliances, such as regulators, etc., should be arranged so that manipulations are quick and easy. Usually the best plan is to place the switches in proper order or sequence, such as 1st, 2d, 3d floors, etc., or Parquet, Balcony, Gallery, etc., in theatres.

42. Dynamo switchboards should be so constructed that all the instruments are in plain sight, and that the dynamo switches can be easily thrown in or out, and should be located as near to the machine as possible.

The location of the switchboards should be such, that the plates or connecting parts on the appliances will not corrode or oxidize, due to the surrounding conditions.

CHAPTER IX.

APPLIANCES AND CONNECTIONS.

43. All connections, on appliances which form a part of the circuit, should be kept clean and bright. The stationary connection should be securely fastened; the sliding or movable connections should have sufficient surface, and bearing tension; and all metal parts should have abundance of carrying capacity so that undue heating will be obviated. The connections should be properly covered and protected. All bases should be of porcelain, slate or similar material, and all parts should be so arranged that access is easy. Switches should preferably be constructed in such a manner that arcing or excessive sparking at the connections is impossible.

44. In joining or splicing wires, care should be taken to have the metal at the point clean where the splice is to be made. The connection must be firm and rigid, and thoroughly soldered and insulated. The splice or joint must be made in such manner that in swinging or bending the wire the solder will not crack or become loosened. In chapter V was shown the relation of the resistance of

the wire to the lamp circuit, and the amount of energy expended in same, but it was assumed that the resistance was equally distributed along the line. Should the line or conductor contain an improperly made joint, that is, one that is loose or corroded, the resistance at that point might exceed the amount of resistance in all the rest of the line. The consequence would be an abnormal heating at the joint, which would gradually extend throughout the whole length of the circuit; the current required for the lamp would be consumed in generating heat, the candle-power of the lamps would decrease, etc.

It is often the case that high grade wire is used, and still the insulation resistance may test low. Generally that is due to the careless manner in which the joint was insulated.

The rule is to have the insulation at the joint or splice equal to that originally on the conductor.

CHAPTER X.

CONVERTER WORK.

45. The converter is practically an induction coil, in which high voltage and a small amount of current are converted or transformed into low voltage and a large amount of current. The general method of constructing induction coils is : a spool having for its core a bundle of fine iron wires ; a few layers of comparatively coarse wire are then wound on the spool, and the ends turned out so that connections with battery or source of current can be made, over the coarse wire. Thoroughly insulated from this are wound a large number of layers of very fine wire, each layer carefully insulated from the others. The coil nearest the core is the primary coil, and the outer, coil of fine wire, is the secondary coil. When the primary coil is connected to a few cells of battery, and the coil is equipped with a rapid make and break device, a powerful electromotive force is created in the secondary coil. The iron core is used for the purpose of increasing the lines of force that pass through the coils, and is composed of a number of fine wires, to avoid the

waste or "Foucault" currents which would be created in a solid core, and which tends to make the coils act sluggishly.

46. In the case of the converter, the coils are constructed in a manner directly opposite to that of the ordinary induction coil. The converter primary consists of a small wire and many layers, and the secondary coil consists of a large wire with a few layers.

The amount of current in amperes supplied to the primary coil is small, but the voltage is high, and the current in amperes created in the secondary coil, is large, but the voltage is low. The size of the coil governs the difference of potential and the current produced, or, in other words, the fewer the number of layers on the secondary, the greater the current in amperes, and lower the voltage.

The converter as applied in practice, is one of the most important discoveries appertaining to the art of electric lighting, and is destined, at an early date, to play an even more important part than it does at present. It is therefore essential that wiremen should familiarize themselves with the workings and construction of the various types of converters. They are usually located on the roofs, or sides of houses, on poles, or in the vaults of structures where the underground system is used.

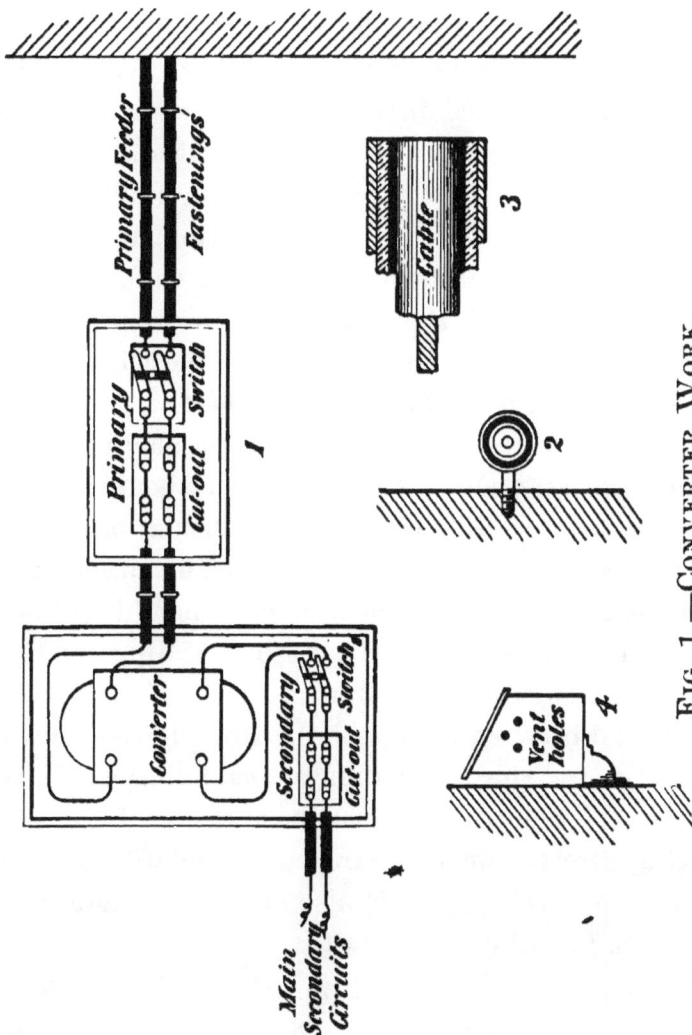

FIG. 1.—CONVERTER WORK.

When placed outside of a structure, a double pole switch and cut-out should be inserted in the line at the point of entrance to the building. When the converter is located in the vault or cellar, the primary wires should be thoroughly insulated and protected from leaks, grounds, short circuits, and mechanical interference or dislocation, and at no point should the conductor, or the material surrounding it, touch the structure. The greatest care should be taken, at the point where the wires enter the building, to exclude gases, moisture, etc. The insulation directly on the conductor should be of the highest grade. The conductor should be inserted in a conduit composed of insulating material, and all should be inserted in a galvanized iron, or such like, pipe. The iron pipe should be fastened to, but kept free from, the building by the use of strong iron arms, having a band of insulating material between them and the iron pipe as shown in Fig. 1. The inner tube or insulating conduits acts as a protection against the abrasion of the covering directly on the wire, and insulates it from the outer iron pipe. It also admits of more freedom in handling the circuit.

A separate conduit and pipe should be provided for each wire, and should extend to a box containing the main cut-out and switch. This box should be fire-proof and water-tight, and should be out of contact with the walls of the structure. Sockets should be provided for the entrance of the pipes, so that the connection between them will be rigid and tight. The box should be of ample size, fitted with a cover or door, which should be kept locked, and fitted with a panel of glass, so that inspection, without exposure, is possible. It should be located as near the point of entrance, and in as dry a place as possible.

The converter should also be located as near to the point of entrance as possible, and should be enclosed in a box, which should be constructed of fire-proof material. The box should be kept free from contact with the building, in a manner similar to the switch-box. It should completely enclose the converter. The roof of the box should extend somewhat over the sides and act as a water shed, and ventilating holes should be cut in the sides. In short, all primary work, which is that part beginning at the point of entrance, up to and including the converters, must be kept free from contact with the walls or floors of the building.

47. The distribution of current is essentially the same as in low tension direct multiple arc systems, and the total loss in the wires must not exceed two per cent. The circuits can either be divided in

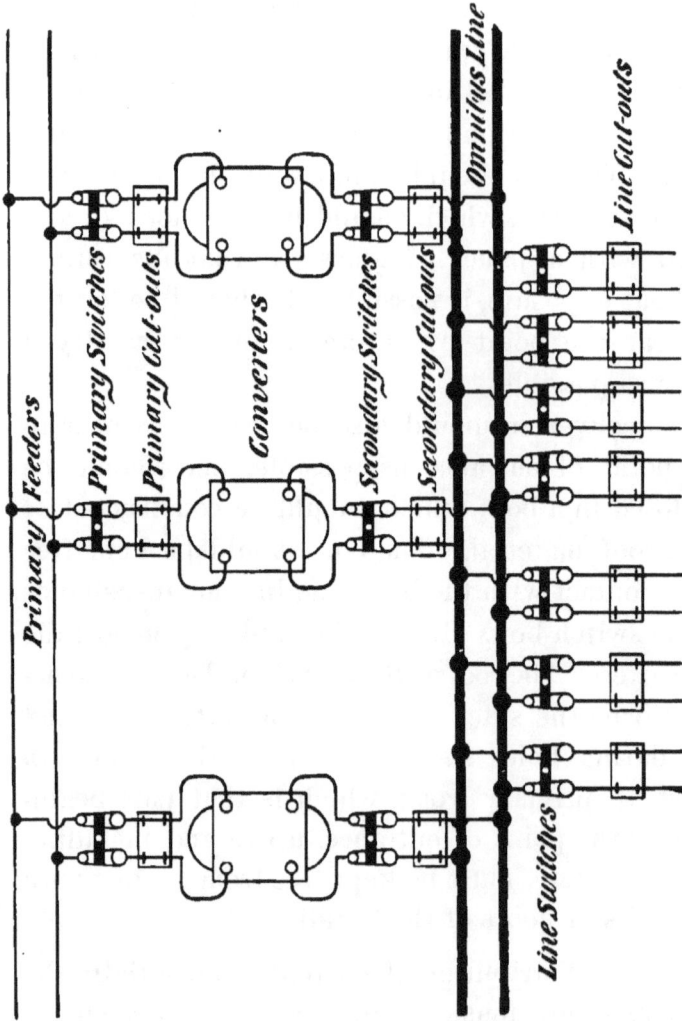

FIG. 2.—CONVERTER WORK.

accordance with the rated capacity of the converter
(providing a separate one for each circuit,) or, the
circuits can be brought to, and connected with
omnibus wires in the usual manner, and also the
secondary of each converter connected to the same,
in a manner similar to the connection of dynamos
in multiple as shown in Fig. 2. Where converters
are to be connected to wiring on the three-wire system two converters are necessary. The primaries are
connected in the usual manner, but the secondaries
are connected in series. The two outer wires connect, one on each terminal, and the middle or neutral wires connect to the wire between the converters as shown in Figs. 3 and 4. Where the larger
sized converters are used, they should be set on a
platform, built up and insulated from the floor;
and the switchboard should be so arranged that
the primary work is separated from the secondary
work. It is preferable to construct two switchboards, the backs facing each other.

In a large converter plant, and where the lighting
is such that the greater portion of lamps are in use
only at stated times, it is the most economical to
arrange the wiring and converters in such manner
that these particular lamp circuits, including the
converters for same, are only connected to the primary circuit at the time of use. The primary and

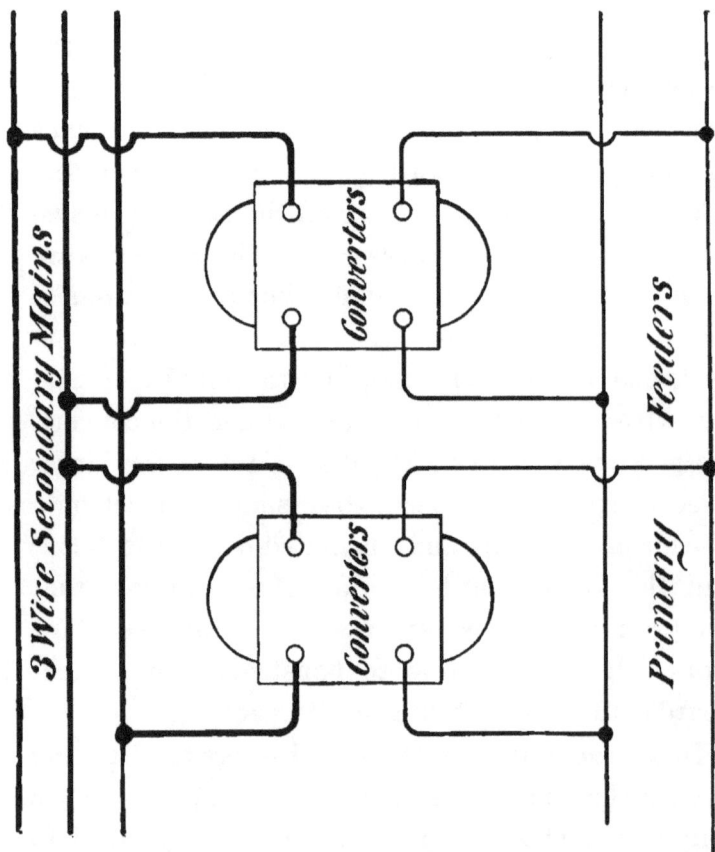

3 Wire Secondary Mains

Converters

Converters

Feeders

Primary

FIG. 3.—CONVERTER WORK.

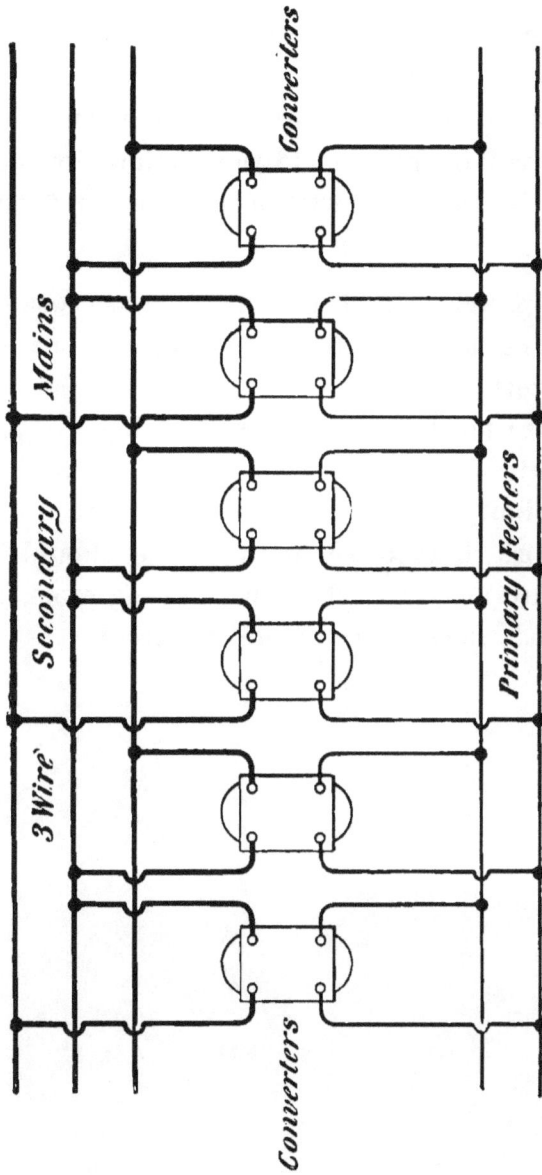

FIG. 4.—CONVERTER WORK.

secondary circuits of each converter should be provided with a double-poled cut-out and switch ; and should more than one converter be connected, the primary circuit should, in addition to those for each converter, be provided with a main cut-out and switch.

48. It is preferable, where converters are located in a building, to have all cut-outs located outside of the converter.

In no case must the two wires forming the primary circuit, be bared at the same time, and in close proximity to each other. In splicing or installing work, only one wire at a time, should be bared. Care must be taken in handling same, so that the body does not complete the circuit.

CHAPTER XI.

OVERHEAD WIRING.

49. It is often necessary in isolated lighting to connect the wires in more than one building to a dynamo, which necessitates pole, or outside work. If the distance between buildings is not great, the best method would be to provide a separate line or feeder for each, and the wires can be fastened to the outside of, and suspended between, the buildings.

It is preferable to use the petticoated glass insulators, instead of the ordinary porcelain knot, for this style of work. They should be kept free from contact with the building, and allowance must be made for swinging, so that the wire will not strike the building. They should be separated sufficiently, *i. e.*, in case of sag they will not come in contact with each other. This also prevents the accumulation of snow or ice. Where the wires enter or depart from a building, generally over a door or window, they must be protected by an extra covering, such as an insulating tube, not only for the purpose of separating and insulating them, but also to act

as a water shedder.   The holes should be bored at
an angle, having the lower point toward the outside.

50. In work of this kind, it is preferable to keep
the two wires forming each circuit, next to each
other in the same manner as the circuits were installed in the interior, so that when the plant is
turned over to the customer, his man, in charge of
the work, will more easily understand the system
of wiring.   The liability of mistakes, when making
changes, will be lessened.

51. In pole line work, the poles should be as
short as possible.   In setting in the ground, the
depth is governed by the nature of soil, the height
of the poles, and the weight of the wire.   Generally,
for a 30-foot pole, 4½ to 5 feet is sufficient.   Should
the ground be swampy, a good plan is to excavate
widely and deeply, to admit placing a strong barrel
in the hole, into which the pole is set.   The barrel
should be filled and packed with small stones and
sand, and the whole covered with the dirt.   This
will form a suitable foundation should the poles be
short and the weight of the wire not abnormal.
The base of the pole should be about 8 inches
in diameter, and the pole should taper somewhat
towards the top.   If the poles are to be longer than

30 feet, a safe rule is to increase the diameter of the pole 1 inch for every increase of 5 feet in length, up to 40 feet; and 2 inches in diameter for every 5 feet increase in length thereafter. The distance between the poles must also be considered. Chestnut wood is generally preferred, although cedar and Norway pine poles result satisfactorily.

52. The cross-arms should be securely bolted to the pole, and about 6 to 8 inches be allowed between wires. Care should be taken when fastening the wires to the insulators, which should be of the petticoated pattern, so that the "tie wires" do not cut the insulation on the wires. It is preferable to provide an extra protecting cover for the wires at all the insulators.

53. The wires should be arranged so that the same wire will be fastened to the corresponding pins on each pole, throughout its entire length. The distance between poles depends upon the weight and size of the wire. A fair average distance would be 200 feet.

CHAPTER XII.

FUSE WIRE.

54. The fuse wire is to an electric lighting system or plant what the safety valve is to a system of steam generation and supply. The wire for fuses is composed of an alloy of tin, lead, etc., and to properly perform its functions should never be connected in a circuit carrying in amperes more or less than the rated carrying capacity of the fuse wire. That is : never use a 10-ampere fuse in a 5-ampere circuit or *vice versa*. Never make any allowance in the carrying capacity of the fuse wire. It has already been tested and standardized by the manufacturers.

55. The fuse wires are connected in the circuit by means of a cut-out block, and when grounds, crosses, or excessive leakage occur, the current increases to such an extent that the conductors become unduly heated, due to their resistance. The nature of the fuse wire being such that it will melt at a lower temperature than the copper wire, it will in the event of an occurrence of this kind, melt, thereby automatically opening or breaking the cir-

cuit. If the fuse wire is too small, it will fuse un-
necessarily, which is a source of annoyance ; but
should the fuse wire be too large, it will only melt
after the load becomes too great, at which time the
line and dynamo are so hot that the insulation is
affected, or the resistance of the copper wire in case
of electrolytic action, will be greater than that of
the fuse wire (making allowance for the difference
of melting points), in which case the wire will melt
and may form an arc. In the event of an occurrence of a short circuit or heavy ground, if no provision in the shape of fuse wires had been made,
the line or armature would melt.

The importance of using only fuse wires, that are
standardized, is evident.

Never assume, unless the fuse wires have been
supplied by the same manufacturer, that because
the cross-section is the same in the case of two fuse
wires, their rated capacity or fusable point is the
same. The fusing point of the wires is governed
by the proportions of the different metals forming
the alloy, and although both of the above-men-
tioned wires may have had sufficient carrying
capacity for the particular circuit, the fusing points
may have been different.

56. In the case of a short-circuit the fuse wire,
even though somewhat larger than necessary, will

melt, but short-circuits is not the only element to guard against. Leaks and grounds are just as liable to occur, and the rating of the fuse wire must be in accordance with these liabilities. To explain the function of the fuse wire more clearly, it may be said that if in a piping system for water, steam, etc., the pipes should be overloaded, they will break or burst at the weakest point. In a wiring system, instead of bursting, the conductor will melt at the point which is weakest. To prevent this happening in the dynamo and conductors, is the function of the fuse wire. Its cross-section or carrying capacity, therefore, is figured only for the current necessary for consumption in the lamps, and should any increase occur, the wire will at once melt and disconnect the circuit. The question of fuse wires should also be considered from a financial standpoint. Should they be too large, and a ground occur, the dynamo is not only generating the current consumed by the lamps, but also that which is wasted through the ground, and should it be in a system in which the dynamo is already heavily taxed, this additional amount which is wasted, will cause an abnormal drop in potential, or still worse, burn out the armature. The only method, whose prompt action can be assured, is the use of fuse wires, the capacity and fusibility of which has been carefully tested and standardized.

57. The insertion of fuse wires in circuits, is of too great importance to be a matter of chance or guess-work, and wiremen should never use fuse wire or links, other than that supplied by the particular company or contractor, in whose employ they are at the time, because the distance between the screws on the cut-outs may be greater or less than for those used by other companies. This will change the carrying capacity of the fuse wires. Or, again, the alloy may be different.

CHAPTER XIII.

INSULATION.

58. Conductors are insulated for the purpose of harnessing or controlling the current, so that it can be directed in a certain defined path, which is generally wire composed of copper. In addition to its inability to transmit current, the insulation should be absolutely moisture-proof, and, if possible, fire-proof.

Rubber, and compounds composed principally of rubber, are, for practical purposes, the best. Rubber is elastic, and can be bent without injury. Being moisture-proof, it acts as a safeguard against the worst element of trouble (see Electrolysis). Although not fire-proof, it will withstand high temperature without melting. This material being also naturally delicate, is protected and strengthened by an outside covering, which is generally composed of fibrous material, treated chemically, so that it becomes to a great extent fire-resisting. The outer covering is not relied on for its insulating qualities, but more, from a mechanical standpoint, as a protection against abrasions, etc. Experience and time has demonstrated that, protection against moisture, is a prevention of fire ; and the best results for interior work, have been obtained when the wires were insulated by a material similar to that mentioned.

59. To secure permanence, the quality of the insulation must be of the best, and absolute continuity must be maintained. The minutest crack, bruise, or pin-hole is sure, sooner or later, to create trouble. It must be handled with the utmost care, and the small amount of extra time spent in careful handling, will more than offset the time and trouble that would be needed to locate and repair insulation damaged by carelessness. The manufacture of high grade wire of this nature, appreciating the importance of the subject, use the utmost care and vigilance in its manufacture. It is one continual round of inspection and testing, from the start to the finish. When the wires are made up in coils, they receive the final test, and if found to be up to the standard, are made ready for market. The same care is exercised in the packing and shipping, and if equal pains were taken after the wire left the manufacturer's hands little trouble would be experienced, due to defective insulation. The greatest source of danger is the rough usage it receives at the hands of inexperienced or careless

wire-men. Another source of trouble is, insufficiently insulated joints or splices. When baring the
wire to make a joint, the insulation should not be
torn off, or cut in a manner similar to whittling
wood. It should be cut at right angles to, and
neatly severed from, the conductor, and then slit
open between the points cut. It will then be an
easy matter to remove the insulation, without strain-
ing it on either side of the exposed part. In solder-
ing, care should be taken to keep the insulation
free from the acid and flame.

60. When the splice is made, it should be insulated in a manner equal to that on any other part of
the line. This is usually accomplished by the use
of a strip of rubber compound, wound or wrapped
around the joint, starting at a short distance back
from the splice, and on the insulation. In wrapping the tape it should be wound at an angle, so
that each of the layers will lap over one-half of the
preceding wraps. This provides a double thickness. If the conditions are adverse, an extra wrapping should be provided, and placed over the first,
in such a manner that the laps of the second covering cross the laps of the first. The heat of the
hand is generally sufficient to cause the material to
adhere and form a solid mass. Over all should be
placed a tape of linen or such material, impregnated

with rubber. When taping splices on exposed
lines in a perpendicular position, the wrapping
should be started at the bottom of the splice and
worked towards the top. In this manner the tape
will act as a shedder.

61. In molding work, the grooves should be sufficiently large to admit the wire easily, without being
forced in. When fastening wires to porcelain insulators with tie wires, an extra wrapping of tape
should be put over the wires at those points, and
care taken, that the tie wires do not bruise or dis-
rupt the insulation. The more recent styles of porcelain insulators obviate the use of tie wires, and
for this reason are a great improvement. Forcing
wires in or around sharp corners is not recommended, and should be done with the greatest care.
When uncoiling wire, be careful not to get a sharp
kink in it. Should a defect in the insulation of a
coil of wire be discovered, the whole coil should
be put aside and subjected to a test before any part
is used. In new buildings especially, the greatest
care should be exercised. Avoid having the wire
either in coils, or strung out, lying on the floor,
while using. The floor is generally covered with
mortar, brick, and other sharp or rough materials ;
mechanics are constantly moving about, and the
fall of a brick, or a barrow wheeled over the wire,

is generally all that is necessary to impair the insulation. The defect may not show at the time, but sooner or later it is bound to be the source of trouble. When passing wires through walls, floors, partitions, etc., a tube of insulating material should first be inserted, so that there may be no danger of tearing the insulation as it is passed through.

62. For pole line or outside work, a strong and durable weather-proof covering should be placed over the insulation.

The quality of insulation can only be determined by tests, and practical application, but to form a fair estimate of the quality simply by inspection, in a general way, it may be said that: the insulation should be tough and lively, that is, it should not resemble leather in the quality of its toughness, but should possess more elasticity, and when stretched and suddenly released, it should quickly assume its previous shape. If when pulled in opposite directions, it parts without stretching, somewhat as dry putty or dough would, it is inferior. The best grades have two layers of insulation, an inner white or red, and an outer black, core, and an additional covering of braid or tape over all. The advantage of the special inner layer is the prevention of the oxidization of the conductor ; and should the outer layer become bruised or impaired, the wire is still protected by the inner layer. The best manner in which to make the superficial tests, as explained, is to cut a small strip of insulation from the wire, divesting it of the tape or braid.

Another method of testing for toughness, is : Taking a piece of No. 14 wire insulated, twist and bend it, until either the wire or the insulation is broken.

In a first-class insulated wire, the copper will break, and the insulation will remain intact.

.

CHAPTER XIV.

ELECTROLYSIS.

63. In discussing this subject, it will be treated only with a view of showing its relation to Electric Light Wiring. While it is really a part of the previous chapter, it was considered too vital to be merely mentioned. It is of the utmost importance that wiremen and others interested in construction work should be familiar with the subject.

Wires are also insulated for the purpose of preventing electro-chemical action. The word "Electrolysis" means Electrical analysis, or analyzing with, or by, the use of electricity. All liquids, with the exception of oils, are fair electrical conductors, and under certain conditions even oil will lose its insulating qualities to a great extent.

Insert the wires forming both sides of a battery circuit, each into a glass tube, closed at one end and nearly filled with water, invert same into a larger jar, into which, water must also be poured. The larger jar should be so arranged that the wires, composed of platinum, will not come in contact with the water in same; the use of the platinum prevents oxidization of the terminals. In completing the circuit, through the water, decomposition of the water ensues. Water being composed of hydrogen and oxygen, it will in short be found that these gases have been generated, the hydrogen in one tube, and the oxygen in the other, and in proportion to their relative quantities. The separation or decomposition is caused by the action of the current when flowing through the water.

Instead of using platinum wire and two tubes and a jar, we will change the experiment somewhat. We will take a jar filled with water into which are placed the ends of two pieces of ordinary copper wire, connected to battery. When the current is turned on, the water completes the circuit. In this case not only is the water decomposed, but the copper also. The wire forming the positive pole will be decomposed and the atoms will be carried through the water to the negative wire. This action in electric lighting is sometimes termed a "partial short-circuit," and if the resistance of the conductor, either liquid or metallic, is lower than that of the fuse wire, the latter will melt and so disconnect the circuit.

64. The most common causes of electrolysis are, defective insulation on the wire ; joints insufficiently insulated, excessive and continuous moisture,

etc. With insufficiently insulated wire located in
or fastened to a damp or moist wall, the amount of
current escaping or leaking may at first be very
small, but the action of the current, and the oxide
from the copper, in a short time render the insulation absolutely worthless. The leak increases, and
the current flowing through the copper wires, which
are immersed in water, will create the chemical
action, as explained. The copper will be decomposed, and dissolves away into the water. Copper
oxidizes rapidly, and the wire at this point, will in
a short time be wholly decomposed. The greater
the current and the lower the resistance of the
moisture, the quicker this will be accomplished.

Where this defect occurs, the action is as follows:
If the resistance of the moisture or liquid between
the poles, is as low as, or lower than, the resistance
of the fusible strip used in the cut-out, the result
will be the same as in a short-circuit; that is, the
fuse will melt. But should the resistance of the
moisture be far greater than that of the fuse wire,
the decomposition of the wire will gradually be accomplished, and as it advances, the heat, and loss
of electrical energy in the conductors increases, the
lights become dimmer, and the carrying capacity
of the wire decreases, to the point where it melts
instead of the fuse wire in the cut-out.

In other instances, the wire will be so decomposed and so thin that at a certain point it is as fine and sharp as an ordinary needle, and a slight vibration

FIGS. 1, 2 and 3.—ELECTROLYSIS.

of the building will cause the wire to part. and an arc to occur.

65. The foregoing proves the importance of properly fusing the circuits, and using only the best grade of materials. The best and most effective method of obviating the danger, is to use conduits, and high grade moisture-proof wires throughout, providing a separate tube for each wire, and excluding moisture at the joints, etc.

CHAPTER XV.

ADVERSE WIRING CONDITIONS.

66. Where adverse conditions exists, such as moisture and gas, the best results have been obtained by stringing the wires, fastened to suitable insulators, on the face of the walls, etc. The insulators should be as few as possible. The fixtures and wires should be kept free from contact with the building. The cut-outs and switches should be located in a clean dry place, and to accomplish this, the length of the circuits must, if necessary, be increased. The fixture, if one is necessary, should consist of a metallic pipe, and the wiring of the fixture must be two heavily insulated wires. The end of the pipe toward the outlet should be closed by a cork, putty, or insulating compound. The socket and lamp should be enclosed in a water-tight globe. Key sockets should never be used, but all circuits be controlled by switches. The fixture should from time to time, be treated to a coat of preservative paint.

In fixtures placed on the outer walls of a building, the stem or pipe should extend through the entire thickness of the **walls**, and the electrical connections be made in the interior of the building.

All shades on outside fixtures should be of the hood pattern, and be provided with an inner shell or cover, so that the socket and lamp connections are protected from the elements.

67. If conduits are resorted to, the same general directions must be followed, and care must be taken to exclude moisture and gas from the interior of the tubes.

Stables.—The chief trouble in stables is due to the ammonia which is generated in the stalls. The moisture and exposure also affect not only the insulation on the wires, but corrode all the metallic parts of the fixture and the line appliances. Therefore the weakest points are at the joints, sockets, cut-outs and switches, and connection between the fixture and circuit wires.

Breweries.—The conditions are somewhat similar to those which exist in stables, and in some portions of the structure, excessive moisture exists at all times. In other portions, the temperature is extremely high, and in still other parts, very low temperature exists. The same care must be taken, as outlined in the general directions, and in the fermenting room, if the conditions are unusually severe, double petticoated glass insulators set on a bridge or collar, suspended from the ceiling should be used, and the wires fastened thereto. The bridge or collar should be treated to at least two coats of preservative, insulating paint. The fixture wire should be of the same size as that on the lamp circuit, and the lamp circuit should be controlled by a switch and cut-out only, located in some suitable place. For the mains and feeders, conduits are recommended. If conduits are used throughout, care must be taken to keep the interior of the tubes free from moisture, gas, etc.

Oil Works, Etc.—If the conditions are such that the oil does not leak through the floor excessively, then a general observance of the succeeding directions is all that is necessary. In some cases the buildings are of the ramshackle sort, and the oil leaks freely through the floors. Where the leakage is thus excessive, the only safe method is to provide a continuous shedder, for the line and fixtures, consisting of tin, heavily coated with noncorrosive paint. The hood or shedder should be suspended from the ceiling, and the conductors be kept free from contact with the building.

Where explosive gases are generated, or where paints, naphtha, and such materials are used or stored, the conductors should be kept free from contact with the building, and be exposed as much as possible, care being taken to provide against the accumulation of gases, etc., at any point. The fixture wire must be connected directly to the lamp-circuit wires, omitting the cut-out. All the circuits must be controlled by a double-pole cut-out and switch only, and the same must be located where the conditions are most favorable.

68. In all such cases as those mentioned, the use of conduits on the mains and feeders is recommended. In no case must wires be moulded or cleated. Either large porcelain knobs or glass insulators must be used.

Insulated wires, of any sort, must never be imbedded directly in plaster; and where alkalies, or acids exist, an extra covering, such as conduit, must be provided.

Avoid placing wires under tiled floors, etc. Floors of this kind are usually cleaned with sulphuric acid, which being absorbed by the tile and cement, attacks the insulation.

In prisons, asylums, etc., the wires, lamps and appliances must be inaccessible for the inmates, but the switches and cut-outs must be located in a place easy of access for those in charge. All the cut-outs and switches controlling particular circuits on the different floors should be located in a place and manner corresponding to the conditions on the other floors. This will tend to simplify matters, and will not confuse the attendant, when it is necessary for some reason to turn on the lights quickly.

CHAPTER XVI.

THEATRE AND STAGE LIGHTING.

69. In theatre wiring, the wires, cut-outs and switches should be located in places out of the reach of inexperienced or malicious persons, but at the same time of easy access to those having charge. The cut-outs and switches should be grouped in as few places as possible, so that the turning on or off of lights may be quickly accomplished. All the lights in the theatre proper and stage should be controlled by switches at the stage switchboard. The circuits controlling lights in the dressing-rooms, cellar, and such places, should also ramify or branch out at this point.

The public lights, that is : the lights on the side-walk, entrances, lobbies, foyers, waiting-rooms, etc., should be connected to switches, all at one point, and controlled from the front of the house, usually in or near the ticket office, or foyer. These circuits should be connected with a separate feeder direct from the source of supply, and independent of the circuits for the lights in the theatre proper.

70. The lights in the auditorium are located in the dome ; proscenium arch ; side-lights in the gallery, balcony, and parquet, and in the private boxes ; the front of, or face of the gallery ; balcony and boxes ; and on the stage. All the circuits for these lights, except the stage lights, are generally connected to the same regulator.

The lights for the orchestra or band, are placed in deep metal shades, painted green on the outside, and white on the inner surface, so that when the lights throughout the house are turned down, or dimmed, the glare of the music lights will not counteract the effect. They must be arranged so that the light is reflected on the music only. The method of distribution is generally as follows : Separate feeders, extending from the switchboard on the stage, are carried to the centre of each circuit, at which point the feeder terminates into a double branch cut-out, and the lamp-circuit wires are carried to the different lamps on the circuit.

A good plan is to provide a separate circuit, connecting with a switch and cut-out at the stage switch-board with the :

Gallery side lights.
Balcony side lights.
Parquet side lights.
Gallery "face" lights.

Balcony "face" lights.

Private box lights, outside of dome lights, generally two circuits ; proscenium arch lights, two circuits. If, in addition to the side lights in the house, chandelier or ceiling lights are placed in the rear of each, a separate circuit for each floor should be provided, and run similarly to the side light circuit.

71. The rate of loss in the conductors is usually computed at 5 per cent., and divided as follows : 1 per cent. loss in the lamp circuit, $1\frac{1}{2}$ per cent. loss in the feeder from stage switch board to the centre of distribution of lamp circuit, and $2\frac{1}{2}$ per cent. loss in the feeders from the dynamo-room or other source of supply. The stage lights generally consist of the foot, borders, bunch, entrance, ground, projecting, and such other lights as may be used for scenic or other effects ; and a few "working" lights.

72. The foot lights generally consist of three circuits, provided with the plain red and green, or blue glass bulbs, respectively. The position and construction of the fixture or reflector, and the lamps, should be such that the lamps are out of the line of sight, and the reflector so placed that the view from the auditorium will be wholly unob-

structed.   The lamp should be placed close to the reflector, and in such a manner that the light is reflected on the stage.   In addition to the electric lights, provision is made for gas lighting, and as the space is limited, the work should be so arranged, that in the event of the gas being lighted, the excessive heat will not affect the electric work. A satisfactory method is to construct the foot lights in a manner similar to that shown in Fig. 1.   The

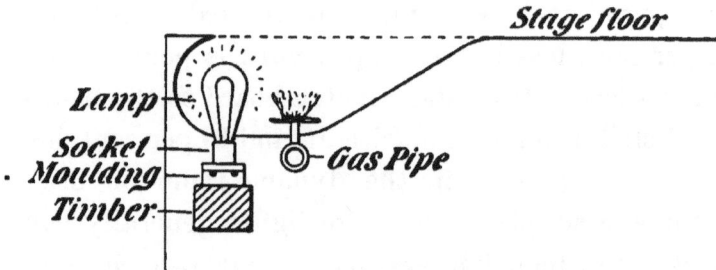

FIG. 1.—THEATRE LIGHTING.

work is located under the stage, where dislocation is improbable, and is securely fastened so that only the bulbs of the lamps are visible.   The work is covered and protected by heavy sheet tin, and holes are cut to correspond with the entrance of the sockets.

This method will allow of placing a large number of lamps in one row.   The three circuits are so arranged that the colors will alternate.

Another method is, placing the wires on the outside of the reflector, having a frame work of wood, and placing the lamps at an angle, as shown in Fig. 2. These circuits are connected to the regu-

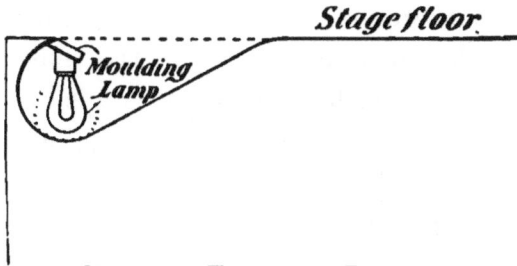

FIG. 2.—THEATRE LIGHTING.

lating device, either separately, or arranged so that one regulator can be used for either, and the number is governed by the width of the opening— usually as many as can be placed without crowding. The inner surface of the foot light reflector should be coated with either white paint or lime.

73. Border lights are rows of lights located in a reflector, and suspended between the curtains in the "flies," and high enough to be out of the sight of the audience; they are used for lighting the body and back of the stage, scenes, etc. The reflector is formed to throw the light in the direction mentioned, and also prevents the light-giving medium from being seen in the front of the house.

Each border usually consists of one row of plain glass bulbs, and the number of rows and lamps on each is governed by the width and depth of the stage, and the number of side entrances. Usually five rows are employed, and the two front, and the three rear borders are connected to separate regulators. At the opening, or front of the reflector is placed a wire screen, to protect the lamps and also acts as a fender in dropping or raising curtains. Each border consists of one or two separate circuits, and the fixture is connected to the feeder by means of flexible cables, the ends of which terminate in connecting plates in the form of a plug, which corresponds with a receptacle in the connecting block, to which the wires from the switch-board are connected. Where two circuits are required, the lamps are connected alternately.

74. Bunch and entrance lights are used for lighting the entrance, or for lighting a particular part of the stage, when the remainder of it is in darkness, etc., and also for general illumination.

The bunch lamps are usually placed on an iron fixture, arranged so that they can be raised or lowered or turned in any direction. The entrance lights are usually lamps placed on a strip of wood in the shape of a row of lights. The connections to the circuit are similar to those of the border lights, a receptacle for connections in the floor being provided on each side of the stage and at each entrance. These pockets or floor receptacles must be fitted in such manner that moisture is excluded, and constructed of fire-resisting material.

The circuit from the switchboard is located on the ceiling under the stage, and is not connected to a regulator.

75. Ground lights or rows, are constructed in a manner similar to the entrance rows, and are used for illuminating the back of hedges, water-falls, platforms, balconies, etc., in the scenery, and are laid upon the floor of the stage, at the point to be illuminated. The receptacles used for the bunch lights are made use of for these. All the stage fixtures are portable, and the arrangements for connecting and disconnecting must admit of being quickly accomplished.

76. Projection lights are used for special purposes, such as throwing a beam of light in the shape of a streak of lightning across the stage, or directing it on a person, or object, or to suggest moving water, etc. The projection light consists of an arc light and suitable resistances, and is specially designed for this class of work. The wires from same can be connected with the incandescent wiring

system used in the theatre. The lamp is provided
with screens, lens, light filters, etc., and special
appurtenances for lightning effects. The fixture is
portable and the reflecting portion is arranged so
that it can be turned in any direction, or raised or ·
lowered. The flexible cable, used for connecting
the lamp with the system, is similar to that used
for border and bunch lights. The connections and
locations are temporary, as the lamp is carried
from one point to another according to its uses.

There is, however, a permanent line or circuit,
carried from the stage switchboard to a connecting
block, located in the front of the gallery. A projecting lamp is used at this point to throw light of
different colors on the ballet, or other quickly
moving objects on the stage.

77. In addition to the lights mentioned, others
of a temporary character, are used for special purposes, such as chandeliers, newel posts, candelabra,
hearth, moon, sun, etc. The fixtures for the first
three are constructed of wood. The hearth lights
represent a fire in the grate, etc., and are usually
ordinary white glass lamps, placed in a receptacle
representing a grate, and the grate is covered with
red paper or mica. The fixture which represents
the moon or sun, is constructed ordinarily of wood,
in cylindrical form, a circuit of red and a circuit of

ordinary lamps are placed therein, and the face of the cylinder is generally of tinted glass, sometimes formed and marked to represent the object. Each circuit is connected with a toning device, so that any tint may be obtained. By the use of the toning device, representations of setting sun or rising moon will be faithfully borne out. The lamps must be located at a distance from the face or glass, so that spots of light or the shape of the carbon will not show.

Nearly all the materials used on the stage for theatrical purposes are flimsy and highly inflammable, and the conditions are such, that a fire once started will gain rapid headway. The work of the stage hands in setting scenes, etc., is done in a hasty manner. It is therefore necessary to provide materials and to locate them in such a manner that they will withstand the existing conditions; and care should be taken in fastening permanent circuits, that they are not on temporary structures, liable to change.

The switchboard should be constructed, as far as possible, of fire-proof materials, and the whole should have a covering similar to the flexible roller top used on desks, lined with asbestos, etc., so that when the switchboard is not in service, it prevents interference with appliances, and also acts as a safeguard against falling objects, water, etc. The switches, cut-outs and regulators should be provided with a suitable name plate, and be so located on the board that connections with the lamp circuits can easily be made to provide for the various combinations necessary in this class of work. At the top of the board should be placed a pilot lamp, from each circuit, to act as a guide for lighting or toning purposes.

CHAPTER XVII. .

PLANS OF DISTRIBUTION.

78. All circuits should be so distributed and connected throughout a structure that practically the pressure or difference of potential will be the same. It is often the case that the total amount of loss is small, but incorrectly divided ; that is, the greater part is allowed in the lamp circuit or sub-feeder, when it ought to have been allowed in the mains or feeder, or *vice versa*.

Assuming 5 per cent. to be allowed in the conductors, it is generally best to distribute the loss as follows : 1 per cent. in the lamp circuit, 1½ per cent. in the mains, and 2½ per cent. in the feeders.

The chance of having all the lamps on any one lamp circuit in use is greater than the chance of having all the lamps connected to a main in use ; and a small loss in the lamp circuit, will allow of placing a few additional lamps on those circuits without appreciably affecting the whole wiring system. The method of distribution is governed by the manner in which the lamps are to be used. In some instances, the lights are in use continuously;

in others a certain proportion are in use at any one time, and in still others, certain sections are used at stated times. The distribution is therefore guided by these conditions. The best results are secured when a separate feeder is provided, for the separate sections having different conditions of lighting.

79. All diagrams given here will show only one wire, unless otherwise mentioned, and may be used for the 2 or 3 wire system.

Fig. 1 represents a cut-out box, either in a panel or closet located in the centre of its lighting area, and shows the connection therewith of the floor mains, which are connected with the vertical mains.

Fig. 2 represents a main and feeder, and more particularly the various methods of distributing from the mains to the different ramifying points or panels. A represents the feeder from the switchboard in the dynamo room, connecting with the mains B at the point X, which is, in this case, the main central point of lighting ; and the lamps are figured as being at a distance equal to that between the switchboard and X. B and D represent the vertical mains, and if the lights are evenly divided on each floor, the distance is equal to one-quarter the total length, B and D connect with floor mains at each floor. On the 4th floor M and M represent

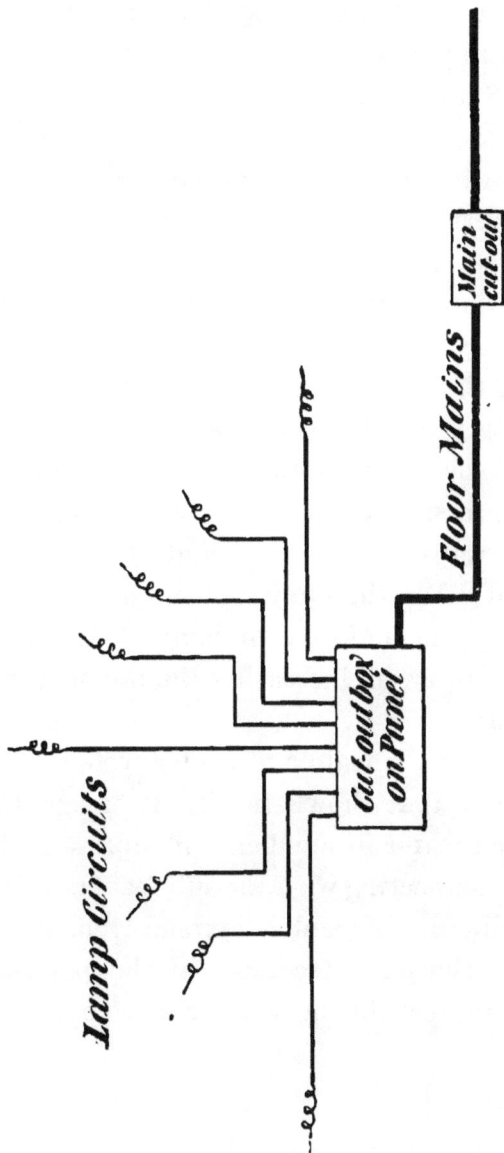

FIG. 1.—PLANS OF DISTRIBUTION.

the floor mains connecting with the cut-outs of the
lamp circuits.   On the 3d floor, to equalize the
pressure, on account of two distributing points, the
floor mains are again divided, as shown.   On the
second floor S and S' each represent a floor main,
and are necessary to equalize the pressure on account of the difference in the number of lights at
each box, and the distance, the box y' being quite
close to the feeding point.

On the 1st floor is shown a method which should
be avoided, because it is impossible to equalize the
pressure.   If the pressure at the lamps from box y'
is normal, the pressure at the lamps from box y
and y^1 would be above normal, and, on the other
hand, if the pressure at the lamps from y is normal
that at the lamps from box y' and y^2 would be
below normal.   We therefore have a choice of two
evils, running the majority of lamps below their
rated candle-power, or, increasing the pressure, and
lamp breakage.

Fig. 3 represents a main and feeder system somewhat similar to that shown in Fig. 2, except that
the floors are greater in number. F represents the
main feeder, connecting with the sub-feeders F^1 and
F^2 at the point A. F^1 and F^2 extend from A, and
connect with the main feeders c at the points D
and E. M represents the floor mains connecting

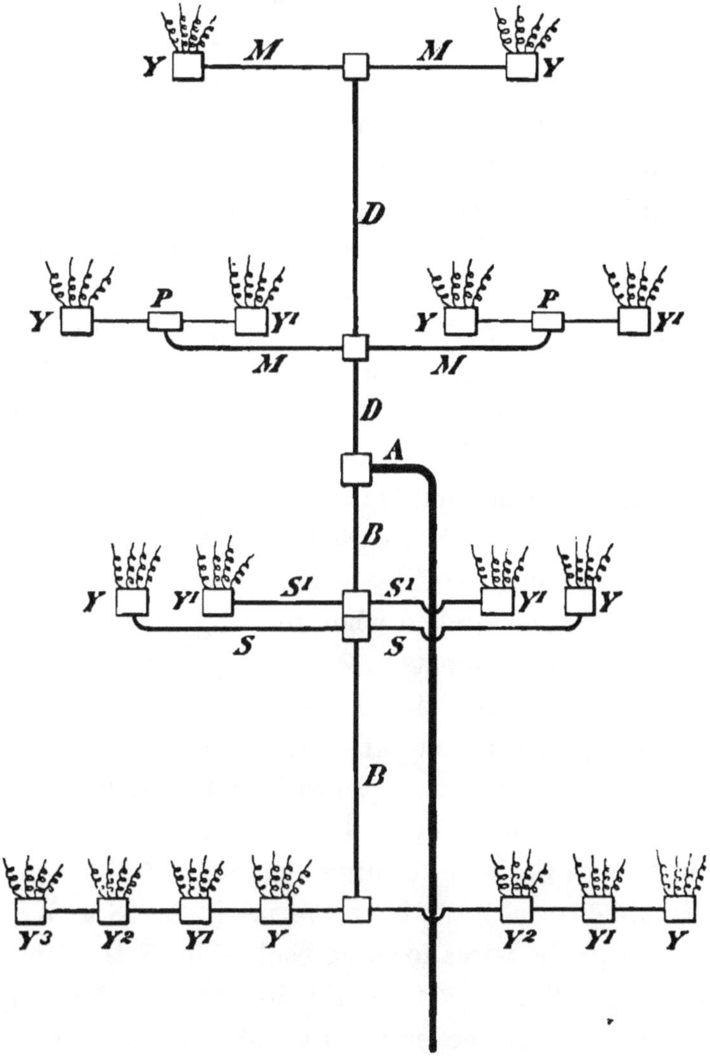

FIG. 2.—PLANS OF DISTRIBUTION.

with the lamp circuits. In large buildings the feeder systems, as shown in Figs. 2 and 3, can be provided, for each side of the building.

Fig. 4 represents a plan of floor lighting, used in factory or general exposed work. A represents the main feeder, connecting with the sub-feeders B and B'; at the point Y the sub-feeders connect with the main wires C and C^1 at the point X, which in turn connects with the lamp circuits L. Fig. 5, represents a system of distribution somewhat similar to that shown in Fig. 4. In Fig. 5 the wires forming both sides of the circuits are shown. This method is made use of where all the lights on any circuit are used at the same time, and are turned on or off by means of switches. This method can be applied when lighting large spaces, such as sheds, depots, etc. A represents a main feeder forming one side of the circuit, and its carrying capacity is computed at the total number of lights, and connects at B with AA, which is the main wire forming one side of the circuit and connects with the wire forming that side of the lamp circuit on each circuit at the point E. S and S^1 represent the wires forming each side of the lamp circuit. C^1 C^2 C^3 C^4 are divisional mains forming one side of the feeder circuit and connecting with S^1 at the points K. It will be seen that the feeder

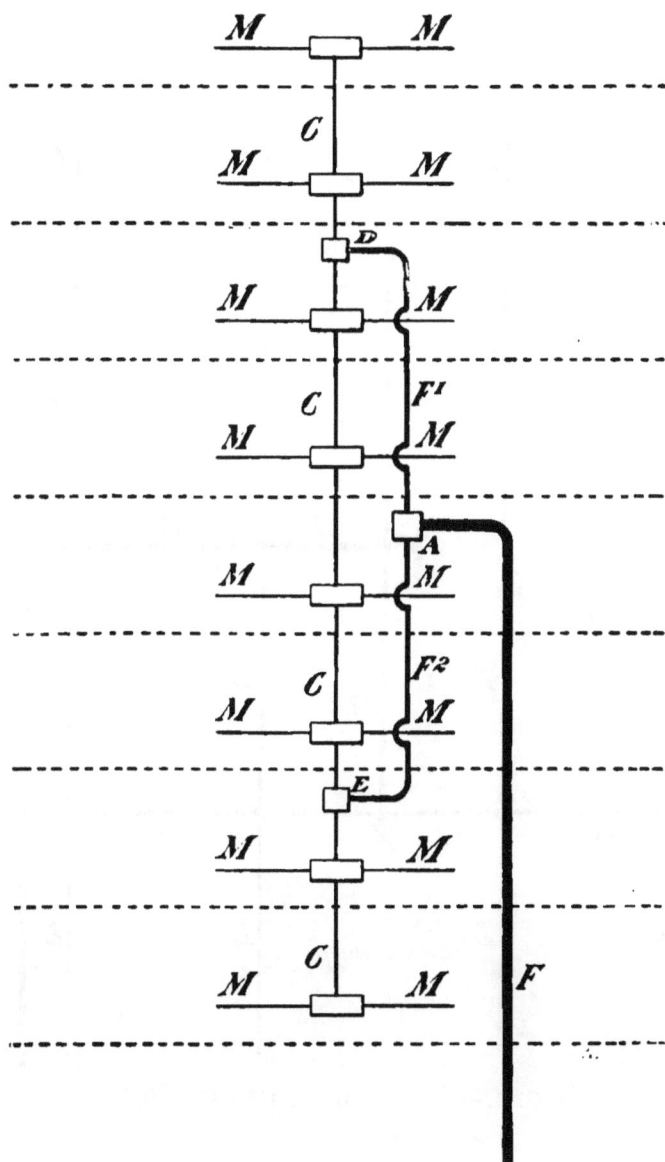

FIG. 3.—PLANS OF DISTRIBUTION.

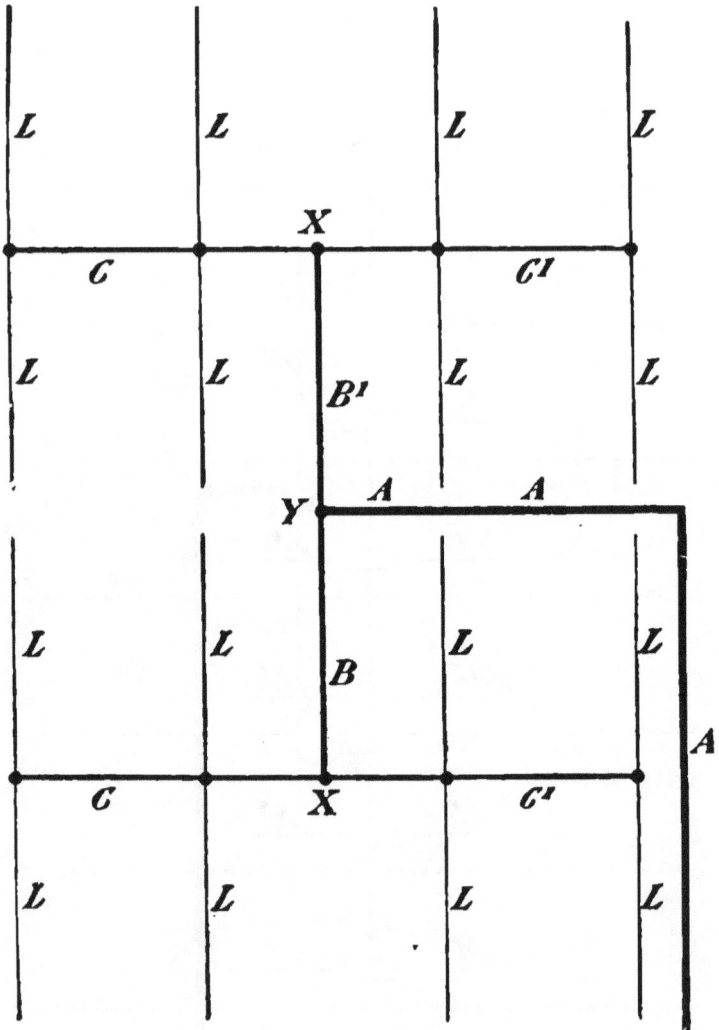

FIG. 4.—PLANS OF DISTRIBUTION.

A, and mains AA and the lamp wires S, form one side of the entire circuit, A and AA being common carriers to all. On the other side of the circuit S^1 forms one side of each lamp circuit independent of each other, and the feeders marked C^1 C^2 C^3 C^4 are connected to S^1 and also independent of each other. At the terminals of the feeders are placed single pole switches, by means of which any of the circuits can be connected without affecting the other.

Fig. 6 represents a system somewhat similar to that shown in Fig. 5, except that in Fig. 5 the feeders were connected at the centre of the length of the mains. In this case, sub-feeders are necessary, due to increased length, so that the pressure at the point X in the mains would equal that at the point Y. The method of connecting the lamps, or throwing them in or out of circuit, is the same as explained in Fig. 5.

Fig. 7 is somewhat similar to Figs. 4 and 5, except in this case two separate feeders are provided for the purpose of equalizing the pressure, and forming two separate circuits of each row of lamps, each circuit being independent of the other, on the same row, and also of those on the other row. One side of the lamp circuit wire is disconnected at the middle, and each half is fed or connected to a separate divisional feeder. The

method of throwing in or out of circuit by switches is similar to that as described in Fig. 4.

Fig. 8 represents a system of distribution applicable to large isolated plants, and by a few alterations in the method can be used for central station distribution. (The diagram shows a "flat" plan). In large hotels the lighting conditions are constantly varying ; that is, all the lights in one section of the building may be in use at one time, and most of those connected on the feeders in the other sections not in use, and *vice versa*. Consequently the pressure on the wires is undergoing variations constantly, the result of which would be excessive lamp breakage, unsatisfactory service, etc. Under certain conditions it is not practical to wire at a low rate of loss, on account of the immense increase in first cost. The interest on the investment will more than offset the small additional cost of pressure equalizers and the lost energy in same.

Therefore, when the conditions exist as stated, the conductors are installed at a higher rate of loss. In the diagram, F^{1} to F^{3} represent the vertical feeders, connecting to the mains and floor mains, in the different sections of the building. The diagram also shows the different methods of feeding, which can be employed in each section. K repre-

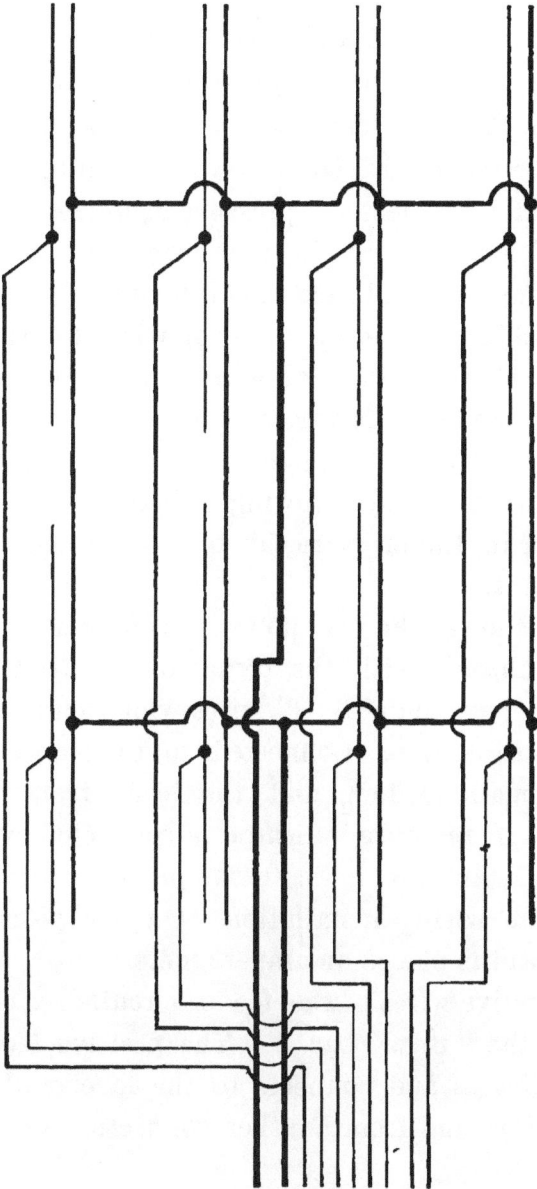

FIG. 7.—PLANS OF DISTRIBUTION.

sents a "crib" to which all the different sectional
feeders connect. E to E^4 represent main feeders
connecting to the "crib" and extending to the
dynamo switchboard. On the wire forming one
side of each is connected a pressure equalizer, connecting with the crib at the point where the main
feeders connect; and, extended to the dynamo
room switchboard, are run a pair of wires for each
feeder, which are connecting with a pressure indicator. As the pressure varies at any of the feeder
points on the "crib," the same will be indicated on
the instrument and by throwing coils of the equalizer, out or in, the pressure can be maintained at a
certain point.

Fig. 9 shows the method of connection, at
the switchboard, with the dynamo and feeders.
A and AA represent "Bus." wires which act as the
main or trunk line. Connected to this, are the
different main feeders, and the leads from the
dynamos. A pressure indicator is connected with
the "Bus." lines, to indicate the pressure in that
part of the wiring installation. On the positive
dynamo lead is placed an ampere-meter.

The negative wires of the feeder circuits are connected to the "Bus." bar A. The positive feeder
wire terminates and connects to the faceboard of
the equalizer, and from another connection on the

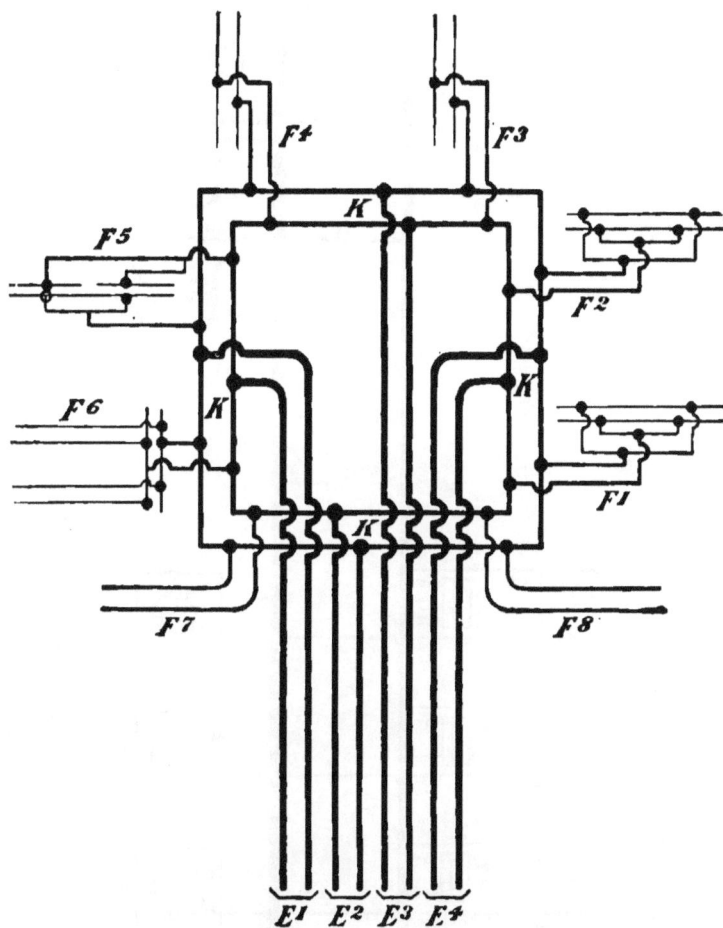

FIG. 8.—PLANS OF DISTRIBUTION.

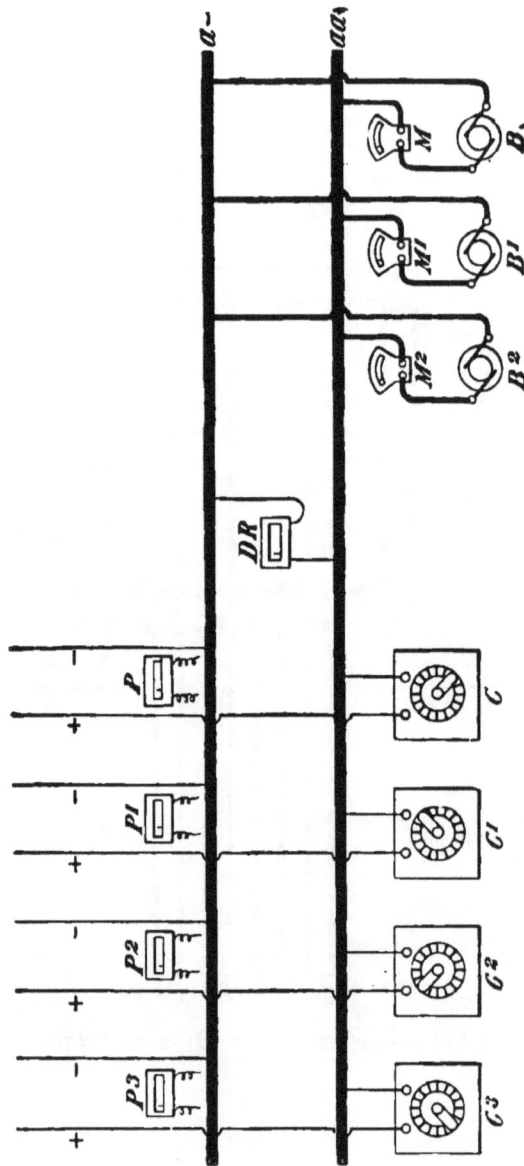

Fig. 9.—Plans of Distribution.

equalizer faceboard; and connecting with *A A* of the " Bus." bar, is run another wire, equal in cross-section to that of the feeder. The equalizer and pressure indicator of each feeder circuit should be placed directly over each other, so that when throwing coils in or out of the circuit, the effect can be noticed. It will also tend to simplify matters.

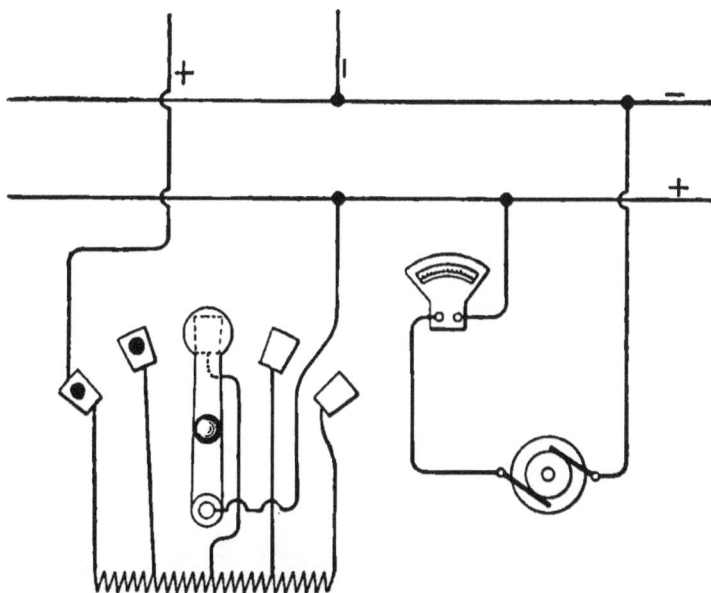

FIG. 10.—PLANS OF DISTRIBUTION.

Fig. 10 represents the coils and method of connections, in the equalizer, and to the feeder, and " Bus " bar or wires on the dynamo switchboard.

Assuming each coil to have a resistance of 1 ohm, and that there were 10 coils, it will, from the connections, readily be seen how the resistance can be varied.

CHAPTER XVIII.

DISTRIBUTION OF LIGHT.

80. In locating lamps, they should be so distributed, that each lamp performs its equal share of the lighting, and is so located that, if the candle power of each lamp is the same, the diffusion of light will be equal, and if the number of lamps is sufficient, a space can be illuminated in such manner that the light is practically equal throughout.

The number of lamps is governed by the space to be lighted, the purpose for which the structure is used, etc.

The practical unit of light is the 16 candle-power lamp, and for ordinary illumination, one 16 candle-power lamp for every 100 square feet, suspended about 8 feet from the floor, is allowed. By 100 square feet is meant a space 10 ft. x 10 ft., or 8 ft. x 12 ft., etc. The amount of light, on this basis, is allowed only when the conditions are such that ordinary illumination is required, as in sheds, depots, walks, etc., and where close inspection of materials, etc., is not necessary. In waiting-rooms,

ferry houses, etc., the allowance is generally one
16 candle-power lamp to every 75 square feet. In
stores, offices, etc., for ordinary lighting purposes,
60 square feet is the usual allowance.

81. Fig. 1 represents the ideal method of distribution, but in practice is objectionable, where
pendants from the ceiling are used. When the
lamps are located close to the ceiling and a fixture

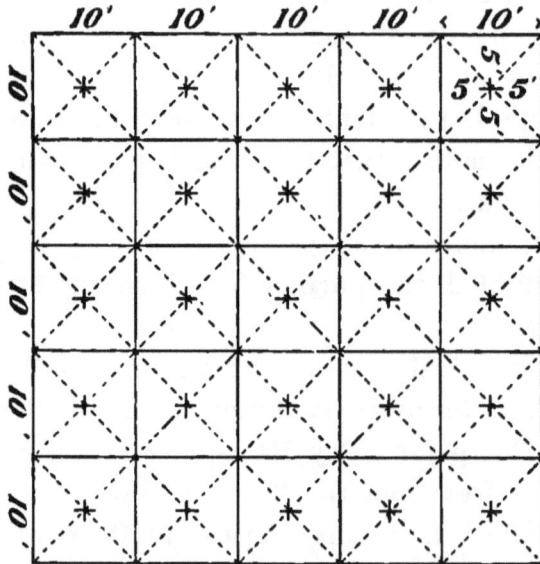

FIG. 1.—DISTRIBUTION OF LIGHT.

or finish, corresponding to the surroundings, is
made, the effect is very good. In the diagram,
assuming the space to be lighted is 50 ft. x 50 ft.,

and the candle-power of the lamp is 16, to find the number of lamps necessary, 50x50 = 2500 ft ÷ 100 = 25—16 candle-power lamps. Dividing the ceiling in squares of 100 ft. or 10 x 10 and placing a lamp in the centre of the square, each light will have the same amount of space to light, and the lamps will be an equal distance from each other.

If in the same space, it is intended to place two lamps in a cluster, to obtain an even diffusion it is

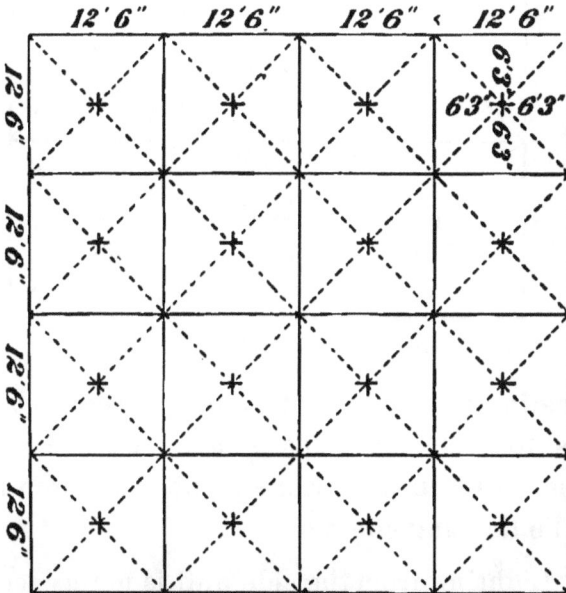

FIG. 2.—DISTRIBUTION OF LIGHT.

necessary to use 32 lamps instead of 25 as in the first case.

Fig. 3 shows a method of distribution, each outlet having a cluster of four lamps instead of single lamps.

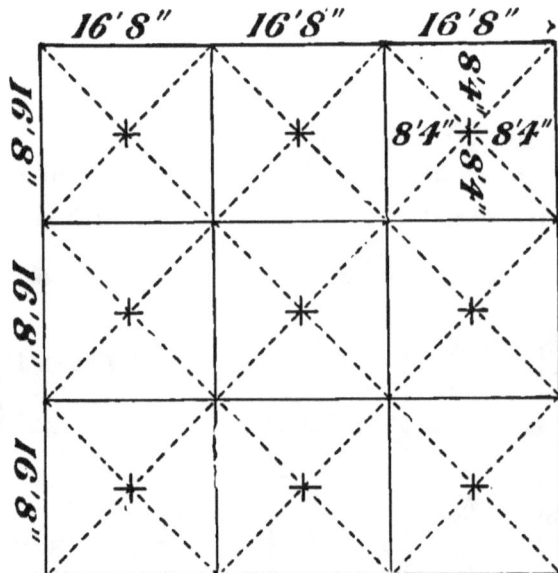

FIG. 3.—DISTRIBUTION OF LIGHT.

Dividing the lamps into clusters, requires more lamps for the same space, and the greater the number of lights at any one point, the more uneven will the lighting effect be.

82. Lighting from the side wall is not as economical as from the ceiling, and the results are limited. Ordinarily side lights are only used to equalize the diffusion of light, when the lighting from chande-

liers, or in a small room, and where a fixture from the ceiling will impart a crowded appearance. For the same reason, it is undesirable to place two chandeliers in the same room. In still other instances the chandeliers would obstruct the view, and the lights would be in the line of sight. The best results, where reflectors are used, are obtained by placing the bulb lengthwise, or parallel with the reflector; and by placing the lamp near to the surface of the reflector, the reflecting power is increased. When placing shades or globes over the lamps, allowance must be made in the amount of light, as a large quantity is absorbed, according to the shape, color, and proximity to the lamp. The surrounding, or prevailing color, of the walls, etc., must also be considered in the location and number of lamps. Where shadows are created, due to pilasters, etc. (where the light is evenly diffused there will be no shadows), to overcome the shadows, locate some lamps on the side of the pilaster, on which the shadow is thrown, or a circle of lights around the pilaster. This will remedy the defect.

83. Lighting by electricity, on account of the absence of extreme heat, admits more artistic display, than does any other artificial illumination. Lamps can be located in recesses in the walls and ceilings, exposing only the lower half of the lamp

bulb ; or, they may be concealed entirely and for their cover may have a finely finished piece of artistic glassware, etc. They can also be arranged to form geometrical figures, or placed in the centre of, or at the intersection of the figures forming the decorations. When placed in recesses and covered by glass, the distance between the lamp and glass cover should be such, that the shape of the carbon or lamp is not visible on or through the glass.

In lighting theatres, concert rooms, lecture halls and similar places, in which there is some objective point, such as a stage or platform, the division, location and quantity of light should be such, that the proper effect at the objective point is secured, and that the light does not strain the eyes of the audience, and that the line of vision is not obstructed by light between the objective point, and the audience. The lights should be so arranged that there is no reflection in the eyes. They should be located at the back of, or considerably above, the audience.

When lighting show windows in stores, etc., locate the lamps as nearly as possible in the corner where the front intersects with the ceiling, providing a tin shade or reflector having a white surface, and extending the same the entire length of the front, and placing the flat of the lamps parallel with the reflector, and quite close to it. The sidewalk should be darkened as much as possible to obtain good results.

To illuminate stained and cathedral glass windows or ordinary-sized windows, form a reflector on each side of the window to be lighted, and place therein a sufficient number of lamps, so that a strong light will be obtained, and locate the same at a distance from the windows, so that the diffusion of light will be equal, and that the spots of light, or the carbons of the lamps, will not be discernible. Increase the dimensions of the window, and the number of lamps and distance between same and the window, must be increased.

84. One of the most difficult problems, in the art of lighting is the illumination of large paintings, etc. In fact, each subject is a separate problem, and must be treated accordingly. The treatment changes according to the size, shape, prevailing colors, if covered with glass or not, and the surroundings. The usual method is to place the lamps in a reflector, which is as long as the width of the picture, the reflector is shaped according to the size of the picture and the number of lights and distance between the reflector and picture to be lighted must be such that the light is evenly diffused over the whole surface, and in such a manner that the

picture can be viewed equally as well from either
side, and so that the view will not be impaired by
counter-reflection.

If the painting is suspended from or fastened to
the side wall, there should be no side lights located
in its near vicinity ; and if placed on a stand
or easel in the centre of the floor, care should be
taken not to have any lights at the back of the
painting, which tends to counteract the effect when
viewing the picture.

Paintings placed in a wooden box and covered
with glass, are the most difficult of all, as the surrounding objects will be reflected in the glass.   To
overcome this requires an abundance of light, and
the best results can only be decided by actual experiment.

It is impossible to define rules governing all conditions of lighting.   The object of this chapter has
been to describe, in a general way, the methods
usually employed, and the amount of light necessary for general illumination, and to suggest the
method usually preferable in special cases.   The
style and methods to be employed can only be
decided by experience.

CHAPTER XIX.

DISTRIBUTION OF LABOR AND HINTS TO FOREMEN.

85. The amount of labor is governed by most of the conditions, as stated in previous chapters, but it can be so divided and directed, that the best results will be obtained, from an executive and financial standpoint.

Let us assume a case in which the plans have been drawn in the drafting-room of the electric lighting company, and turned over to the foreman in charge, with instructions to install the work. Let us also assume (which is very often the case), that the draftsman had not seen the building. The foreman, from these plans, estimates the amount of material required, and after ordering same, he should familiarize himself with the structure ; observe which portions of the building are in a more advanced state of construction, and the manner in which the labor in the various branches of the building trades, is distributed.

The plans instruct him as to the general location of wires, and the distribution of current, but he is responsible for the mechanical details, and waste (if any) of labor.   Before the materials are received, he should select a safe place for same, and arrange to keep the "lockup" *locked*.

When acquainted with the conditions of the building, he decides upon the number of men, that can be, to good advantage, employed on the work at any one time.   The work should be started in that portion of the building, which is in a more advanced state of construction, so that, when once started, it can be completed without interruption or delay.   It is advisable to keep the same men in the same part on the different floors of the building, so that they will become familiar with the circuits in that section, as nearly all the office buildings, etc., from the second floor up, are the same.   This will enable the man, after completing one floor, to do the work more quickly on each succeeding floor.   It also demonstrates whether the amount of work is satisfactory, and when testing the result will show whether the man is careful or not.   It is to the foreman's interest to acquaint himself with the capabilities of the men under his supervision, so that he is better enabled to decide which to place on the more exacting and difficult work.   The lengths of the circuits as shown on the plan should be conformed with, as nearly as possible.   The exact sizes of wires as noted on the plan should be used in all cases, but, should it become necessary to change the work, from the "lay-out" on the plan, the company should be notified in season, so that instructions regarding the change may be given, and that the work will not be delayed, waiting for instructions.

86. When the top or lamp circuits are nearly completed, provision should be made for testing them. It is also time to arrange for the cut-out boards which are to be set in the panel or closet. About this time, also, the question of running the mains and feeders should demand attention.

The foreman should continually stroll through the building, watching the work already completed; the work under way ; observing the advancement in construction of the different portions of the building, and continually calculating as to the method of procedure on his own task, so that the wiring work will not delay the work in other branches, or *vice versa.*

One familiar with the method of testing should be appointed to do this work, and his duty should not only be to test, but he should locate and repair faults, and "pick up" and finish the innumerable "odds and ends" which were left undone. One of the most important labor-saving methods, is to "lay-out" and work all the circuits, which run parallel to each other, at the same time, all of which connect to different cut-outs in the same box. It will obviate crosses, and to a great extent simplify matters. It is easier to trace the different circuits.

The same plan is suggested when wires pass through walls, etc., provision can be made, at one time, for all. The most experienced and careful men should construct the cut-out and switchboards, and run the mains and feeder.

87. The foreman should provide himself with a note-book, he will find it to be of service both to himself and employer. He should keep a record of, the date, amount, and kind of material; the result of each test; the changes (if any) in the wiring; additions or omissions in the lamps, switches, etc., the cause thereof; extra work (if any) and by whose authority; the general condition of the plant when turned over to the customer; whether dynamo is belted direct to engine; if so, whether engine does other work besides driving dynamo. If belted to pulley countershaft, note variations of speed, if any, and such other data as may be of interest to his employer, and obviate trouble for the future consumer.

CHAPTER XX.

Preliminary to Rules, Electrical Data, Etc.

88. $+$ is the sign of addition, and is called plus. Thus $2 + 6$ indicates that 6 is to be added to 2 and is read 2 plus 6 or $A + B$, etc.

$-$ is the sign of subtraction, and is called minus. Thus $6 - 2$ indicates that 2 is to be subtracted from 6, and is read, 6 minus 2 or $b - a$, etc.

$\times$ is the sign of multiplication, and is read, times, or multiplied by. Thus 2×6 indicates that 2 is to be multiplied by 6, and is read 6 times 2, and, $a \times b$ denotes that multiplication of $a \times b$.

$\div$ is the sign of division, and is read, divided by. Thus $6 \div 2$ is an indication that 6 is to be divided by 2.

Division is also indicated by writing the dividend above and the divisor below a short horizontal line as in a fraction, thus $\dfrac{6}{2}$ The exponent of a quantity is the number which indicates how often the quantity is used as a factor. Thus, A^3 indicates that A is to be used as a factor three times, and A^3 is the same as $A \times A \times A$.

A' is read, "A square" or "A second power,"
A³ is read "A cube" or "A third power," and, if
A = 6, A' = 6 × 6 = 36, or A³ = 6 × 6 × 6 or
6 × 6 = 36 × 6 = A = 218.

The co-efficient is a number written before a
quantity to show how many times the quantity is
to be taken. Thus 3A would show that A is to be
taken three times, and if A = 6, 3A = 6 × 3 =
3A = 18.

An algebraic expression, is the expression of a
quantity, by means of algebraic symbols ; the
symbols indicate the relation of quantities.

A formula is a method of expressing in a simple
and concise form, a rule or principle, and show the
relation of one to another. Volt is the unit of
electrical pressure, (electromotive force) and is
represented by E.

Ampere is that current having an electric pressure of 1 volt which flows through a wire having a
resistance of one ohm, and is represented by C.

Ohm is that amount of resistance which, in a
conductor, would limit the current, having a pressure of one volt, flowing through same, to one
ampere, and is represented by R.

According to Ohm's law, the current in amperes
is equal to the electromotive force in volts divided
by the resistance in ohms, thus, $C = \dfrac{E}{R}$.

The electromotive force in volts is equal to the product of the current in amperes, and the resistance in ohms. Thus, $E = C \times R$. The resistance in ohms is equal to the electromotive force in volts divided by the current in amperes. Thus $R = \dfrac{E}{C}$.

From these, can be ascertained the current in amperes, when the electromotive force in volts, and the resistance in ohms is known, and,

The electromotive force in volts, when the current in amperes, and the resistance in ohms is known, and,

The resistance in ohms, when the current in amperes, and the electromotive force in volts is known.

Therefore in a machine, the resistance of which $= .5$ ohm and the electromotive force in volts $= 100$, the current in amperes equals $\dfrac{E = 100 \text{ volts}}{R = .5 \text{ ohm}} = C = 200$ amperes, and, $C \times R = E$. 200 amperes $\times .5$ ohm $= 100$ volts, and, $\dfrac{E = 100 \text{ volts}}{C = 200 \text{ amperes}} = R = .5$ ohm.

89. To ascertain the available current in amperes, not generated at the dynamo, but for consumption by or in the lamps, the resistance of the conductors must be added to the internal resistance of the

dynamo, and if the resistance of the conductors equals .0555 ohm, the available current equals,

$$\frac{E = 100 \text{ volts}}{R = .5 \text{ ohm} \times R = .0555} = \frac{100}{.5555} = C =$$

180 amperes.

To ascertain the combined resistance of a number of lamps connected in multiple, divide the resistance of one lamp by the number of lamps. Thus, the combined resistance of 40 lamps connected in multiple, the resistance of one equaling 200 ohms, would be

$$\frac{200 \text{ ohms} = R \text{ of 1 lamp}}{40 \text{ lamps}} = 5 \text{ ohms.}$$

To ascertain the combined resistance of a number of lamps connected in series, multiply the resistance of one lamp, by the number of lamps. Thus, the combined resistance of 10 lamps connected in series, the resistance of one lamp equals 200 ohms, would be,

R of 1 lamp × number of lamps.   Total resistance
    200   ×   10   =   2000 ohms.

To ascertain the combined resistance of a number of series of lamps, the series connected in multiple, divide the resistance of one series by the number of series. Thus, the combined resistance of 4 series connected in multiple, the resistance of one series equal 2000 ohms, would be,

$$\frac{2000 \text{ ohms} = R \text{ of 1 series}}{4 \text{ series}} = 500 \text{ ohms.}$$

Therefore, to ascertain the resistance required in the conductors, so that the loss in same equals, in the desired per cent., a certain proportion of the total resistance, it is necessary, if connected in multiple or series, to first ascertain the combined resistance of the lamps, and assuming, 20 lamps connected in multiple, the resistance of each lamp being 180 ohms, 5 per cent. loss of electrical energy to be allowed, in the conductors; the distance being 350 feet from the dynamo, therefore,

$$\frac{\text{R of one lamp} = 180 \text{ ohms}}{\text{Number of lamps} = \quad 20} = 9 = \text{Total resist. of all lamps,}$$

and, 9 ohms representing 95 per cent. of the total resistance of the circuit at 5 per cent. loss, the total resistance is, $\dfrac{9 \times 100}{95} = 9{,}474$ ohms $=$ total resistance of lamp circuit. The length of the wire being $350 \times 2 = 700$ feet, and, .474 ohm is the resistance necessary to be contained in a wire 700 feet in length, to supply current to 20 lamps located at a distance of 350 feet from the dynamo. If the lamps are connected in series, first ascertain the total resistance of the series, and the method of ascertaining the resistance of the conductor is similar to that, when the lamps are connected in multiple, assuming the same number of lamps, per cent. of loss and distance from the dynamo, to be the

same as in last example, but, that the lamps will
be connected in series instead of multiple, therefore the combined resistance of all the lamps equals,

180 = R of one lamp

20 = Number of lamps

3600 = Total R of all lamps.

5 per cent. being the desired amount of loss in the
conductors, therefore 3600 ohms represents 95 per
cent. of the total resistance of the lamp circuit, and,

$$\frac{3600 \times 100}{95} = 3789.4 \text{ ohms} = \text{total R of lamp circuit,}$$

therefore, 3600 ohms = R of Lamps = 95 per cent.
of circuit, and, 189.4 ohms = R of conductors = 5
per cent. of circuit resistance.

90. To ascertain the resistance of a one or a number of corresponding wires, the method is similar
to that for lamps. Mil. = .001 of an inch, and
when made use of in relation to wiring, is the unit
of length, when measuring the diameter, or cross-
section of wires. Circular Mil is the unit of area
employed in measuring the areas of cross-sections
of wire.

The diameter or cross-section of a wire is
expressed in mils, and the area of cross-section in
circular mils, therefore a wire, the diameter of
which equals ¼ inch = 250 mils, and to ascertain
the circular mils it is necessary "to square" the

diameter, d', 250 × 250 = 62500 circular mils. Foot-mil, equals a wire, the length of which equals one foot, and the diameter, one mil; it is used in practice as a basis for computing the resistance of any given wire, and if the copper is commercially pure (usually 96 per cent. conductivity), the resistance of same, at 75° Fahrenheit, equals 10.79 ohms.

The resistance of a copper wire is equal to its length in feet, multiplied by the resistance of one foot-mil, 10.79, and divided by the circular mils, or "the square" of its diameter, therefore,

$\dfrac{L \times 10.79}{d^2 \text{ or C.M.}} = R$, and, assuming the length to be 1500 feet and the circular mils = 10.381, the resistance would equal $\dfrac{1500 \times 10.79}{10.381} - 1.559$ ohms.

The cross-section of a copper wire, in circular mils, is found by multiplying the resistance of a foot-mil (10.79) by its length (L) in feet, and dividing the result by its resistance (R) in ohms; therefore,

$\dfrac{10.79 \times L}{R} = d^2$ or c. m., and, the resistance, and length of the wire being the same as in last example, the cross-section in circular mils would equal $\dfrac{10.79 \times 1500}{1.559} = 10381 = $ c. m.

91. Having ascertained the resistance of the conductors, at the desired percentage of loss, for any
given number of lamps, the cross-sections of same
can be found, as shown in the previous example.
To demonstrate: assuming 35 lamps, the R of one
= 90 ohms, located at a distance of 160 feet from
the dynamo, and 5 per cent. being the loss of
energy desired in the conductors,

$$\frac{90 \text{ ohms R of lamps}}{35 \text{ lamps}} = 2.43 \text{ ohms} = \text{Total R of}$$

lamps.   2.43 ohms = 95 per cent. of total resistance
of circuit, therefore the total resistance of circuit
equals, $\dfrac{2.43 \times 100}{95} = 2.558 \text{ ohms} = \text{Total R of cir-}$
cuit, and,   2.43 ohms = Total R of lamps,
          therefore,   .128 ohms = R of conductors.
The distance being 160 feet, the total length of the
wire equals 320 feet, and the cross-section in circular mils, of a wire that length and resistance, is,
$\dfrac{10.79 \times 320}{.128} = 26.967$ c. m. = No. 6 B. & S. gauge
wire.

92. The term "difference of potential" denotes
that portion of the electromotive force which exists,
at, or between, any two points in a circuit, and
equals the electromotive force, (in a dynamo) at
the point where the armature "cuts" the lines of

force, minus the amount lost in transmission, due to the resistance of the conductor : To illustrate : In an armature, there is created or generated a certain amount of current at a certain pressure ; the pressure decreases according to the resistance of the conductors ; therefore, the electrical pressure at the brushes is less, due to the resistance of the armature coils, than in the armature, and less in the feeders, than at the brushes, etc.

In the diagram A represents an armature of a dynamo, B represents the commutator or brushes, C represents the feeder, D represents the main, and E represents the terminals at the lamp. At, or in A is "the electromotive force ; that is, the point at which the electrical pressure is the greatest. The pressure at B is less than that at A, due to the resistance of the armature coils, consequently its potency is less ; that is, there is at this point a "difference of potential," which is governed by the

FIG. 1.

resistances of the conductors, and the pressure at B plus the amount lost in the armature coils, equals "the electromotive force." The difference of potential in C the feeder, D the mains, and E at the lamp terminals is governed according to the same conditions. Assuming the electromotive force in A equals 125 volts, and the loss or drop of electrical pressure in the armature equals 1 per cent., the "potential difference" at the brushes B would equal 123¾ volts, and, if the loss in the feeder C equals 5 per cent. of that in B, the difference of potential in C would equal 117½ volts ; and, if in the main wire D, the loss is equal to 3 per cent. of the pressure in C, the difference of potential in same would equal 114 volts ; and, if the loss in the lamp wires, is equal to 2 per cent. of the pressure in D, the difference of potential at the lamp terminals would equal 111¾ volts.

93. The term "electromotive force," is generally used to denote the pressure at the highest pressure point ; in all other parts of the circuit, the pressure is noted as the potential difference, or difference of potential.

CHAPTER XXI.

RULES FOR ASCERTAINING REQUIRED SIZES OF WIRE.

94. The following rules enable us to determine the size of wire necessary, for any number of lamps, at any distance (see safe-carrying capacity), and at any desired loss, expressed in circular mils :

RULE 1. Multiply the resistance of one foot-mil by twice the distance, and by the number of lamps, and by 100, minus the per cent. loss, and divide the result by the resistance of one lamp multiplied by the percentage of loss. The loss should be expressed as a whole number. Thus,

$$\frac{\text{R of 1 Ft.-mil} \times 2 \times D \times \text{No. of lamps}}{\text{R of one lamp}} \times \frac{100 - \text{loss}}{\% \text{ loss}}$$
$$= \text{C. mils.}$$

Example : 80 lamps, located at a distance of 140 feet, R of Lamp = 200 ohms, to determine the size of wire at 5 per cent. loss :

Foot-mil = 10.79 ohms.

$$\frac{10.79 \times 2 \times 140 \times 80}{200} \times \frac{100-5}{5} =$$

$$\frac{241696}{200} \times \frac{95}{5} = \frac{22961120}{1000} = 22961 \text{ cir. mils} = \text{No. 6}$$

B. & S. G. wire.

RULE 2. To determine the size of wire, the loss or "drop" expressed in volts, multiply the resistance of one foot-mil, by twice the distance, and by the number of amperes, and divide the result by the number of volts to be "lost," thus,

$$\frac{\text{R of 1 Ft.-mil} \times 2 \times \text{D} \times \text{amperes}}{\text{Volts drop.}} = \text{c. mils.}$$

Example : 185 100-volt lamps, located at a distance of 260 feet.

The resistance of each lamp = 140 ohms. Drop = 3 volts.

It is necessary to find the amount of current, in amperes, required for each lamp, and according to Ohm's law, $C = \dfrac{E}{R}$ therefore, the current required for, or consumed by each lamp equals,

$$\frac{\text{Volts 100}}{\text{ohms 140}} = \frac{5}{7} \text{ ampere, and,}$$

$$\frac{5}{7} \times 185 = 132.14 \text{ amperes, and,}$$

$$\frac{10.79 \times 2 \times 260 \times 132.14}{3} = 247123 \text{ c. mils.}$$

95. Where the conditions of the lamps do not change, that is, where the wiring is installed for connection, and use of the same kind of lamps, continuously, it will be found handy to have "a constant" already calculated for the different losses,

and it is found by multiplying the resistance of one foot-mil, by two, and the result, by 100 minus the desired loss, and divide the result by the resistance of one lamp multiplied by the loss. The loss to be expressed in whole numbers, thus,

$$\frac{R \text{ of Ft.-mil} \times 2}{R \text{ of lamp}} \times \frac{100 - loss}{loss} = \text{constant, and,}$$

assuming the R of lamp = 200 ohms and the loss = 5 per cent., the constant would be,

$$\frac{10.79 \times 2}{200} \times \frac{100 - 5}{5} =$$

$$\frac{21.58}{200} \times \frac{95}{5} = \frac{2050}{1000} = \text{constant } 2.05 = \text{and, to}$$

ascertain the size of wire, expressed in circular mils, for a given number of lamps, a certain distance, at any desired loss, multiply the number of lamps, by the distance in feet, and the result, by the constant for the loss desired ; thus,

No. lamps × distance × constant = cir. mils.

Example : The resistance of lamp, and per cent. loss, as in previous example ; determine the size of wire required, for 175 lamps, at a distance of 135 feet. 175 × 135 × 2.05 = cir. mils = 48431 = No. 3 B. & S. G. wire.

96. The cross-section of wire should be such, that it will conduct the current without becoming heated to the point where the temperature is greatly

in excess of that of the surrounding air. The evidence of this condition can be got by grasping the wire with the bare hand. All wires become heated when a current of electricity is passed through them ; and by increasing the amount, in amperes, in any given wire, the heat is increased.

According to the rules for wiring, the result is the same in circular mils, for 100 lights one foot as for one light 100 feet. It is easily understood that this is not correct, therefore care must be taken, in short distances to provide a sufficiently large wire, in diameter, so that same will not become unduly heated. The accompanying table is a safe practical guide, and when figuring wire, should the circular mils, per ampere, be less than mentioned in the table, it is advisable to determine the sizes required allowing per ampere, the number as stated therein, rather than by the rules. Allowance in circular mils per ampere, being the safe carrying capacity.

SAFE CARRYING CAPACITY.

(AS ORDERED BY THE BOARD OF FIRE UNDERWRITERS.)

BROWN & SHARPE.		BIRMINGHAM.		EDISON STANDARD.	
GAUGE No.	AMPERES.	GAUGE No.	AMPERES.	GAUGE No	AMPERES.
0000	175	0000	175	200	175
000	145	000	150	180	160
00	120	00	130	140	135
0	100	0	110	110	110
1	95	1	95	90	95
2	70	2	85	80	85
3	60	3	75	65	75
4	50	4	65	55	65
5	45	5	60	50	60
6	35	6	50	40	50
7	30	7	45	30	40
8	25	8	35	25	35
10	20	10	30	20	30
12	15	12	20	12	20
14	10	14	15	8	15
16	5	16	10	5	10
18	3	18	5	3	5
		20	3	2	3

CHAPTER XXII.

ENERGY—POWER.

97. The energy which is developed in a circuit, when a current of one ampere flows through a conductor whose resistance is one ohm, is termed a watt. The watt is the electrical unit of power, and equals $\frac{1}{746}$ horse-power, or, horse-power = 33,000 foot-lbs., that is, a horse-power equals that power which will raise 33,000 lbs. a distance of one foot, in one minute of time, and $\frac{1}{746}$ H. P. = 44.25 foot-lbs. per minute. The number of watts developed in a circuit is determined by multiplying the amperes by the volts,

Amperes × volts, or C × E = watts, or C' × R = watts.

And if expressed in the terms of the mechanical unit (horse-power)

$\frac{C E}{746}$ or $\frac{C' R}{746}$ = horse-power.

If expressed in foot-pounds, C × E × 44.25, or, C' × R × 44.25 = foot-lbs.

Example : Determine the electrical energy, expressed in horse-power, developed in a dynamo, connected to 750 16-c. p. lamps, each lamp requiring ½ ampere of current at a pressure of 100 volts. The loss in the wires equals 5 per cent.

750 lamps × ½-ampere = 375 amperes,

$$C \times E = \text{watts}$$

and, 375 amperes × 100 volts = watts = 37500 and, watts × 44.25 = foot-lbs.

37500 × 44.25 = 1658375 = the energy, expressed in foot-lbs., expended in the lamps, and represents 95 per cent. of the total as the remaining 5 per cent. is expended or lost in the conductors, therefore 1658375 = 95 per cent. of total energy, and

$$\frac{87284 = 5 \text{ `` `` `` ``}}{1745659} = \text{Total energy, expressed in foot-lbs.,}$$

expended in the lamps and conductors, and foot-pounds, divided by 33000 equals horse-power, therefore, 1745659 ÷ 33000 = 53 horse-power.

To find the actual horse-power expended at the pulley of the dynamo, assuming the efficiency of the dynamo to be at 90 per cent., and according to the last example, 1745659 foot-pounds were expended in the lamps and conductors, therefore, the efficiency of the dynamo being 90 per cent., it represents only 90 per cent. of the power expended at the pulley of dynamo, and as 1745659 = 90 %,

and,   193962 = 10 %

1939621 = Total

energy in foot-pounds ; therefore, 1939621 ÷ 33000
= 58.777 = total amount of energy in horse-power
expended at the pulley of the dynamo, and if
belted to engine, 10 per cent. being lost in the
transmission of power, then

58.777 h. p. = 90 %, and

6.53 h. p. = 10 % loss

65.307 h. p. represents the amount of
horse-power at the pulley of the engine.

In the first example it was shown that 87.284 foot
pounds were expended in the conductors due to
their resistance, which equals, 87284 ÷ 33000 =
2.645 or 2¾ horse-power, and the cost, in dollars and
cents, equals 2¾ × lbs. of coal per h. p. hour ×
cost of lbs. of coal

= $ and cts.

This is but another example, of the importance
of carefully considering the question of "loss" in
conductors.

CHAPTER XXIII.

DYNAMOS AND MOTORS.

98. In setting, connecting, or running a dynamo, or motor, unless you are familiar with all its parts, method of winding, and connections, it is very important that a blue print or diagram, showing the different parts, and method of connections, etc., be procured.

99. The machine should be located in a clean, dry place, and where the temperature is not high; it should be isolated, as much as possible, from other machinery, especially in saw mills, machine shops, etc., where more or less dust, or metal filings are flying, but at the same time it should be accessible. If the machine is set on an ordinary floor, provision should be made against vibration. The base frame should be treated to a coating of hot paraffine, for closing the pores, and a thick coating of shellac. In putting the machine together, care must be taken to have all the parts clean, and all connections, bearings, etc., must be put together so that the "fit" will be perfect. All parts, especially the armature and magnets, must be handled with

the utmost care, and must not be handled any more than is necessary.

100. Before starting, adjust and test the brushes for tension. See that the main connections in the circuit are open ; examine all connections ; provide each cup or automatic lubricator, with sufficient oil, and see that the feed is in proper working order.

In a shunt-wound machine, when up to speed, the brushes can be dropped on the commutator, and if hand regulator is used, the resistance should be thrown out of the field, until the needle on the indicator is at the point showing the required pressure. The lamp at the head of the dynamo is generally a guide.

101. It is preferable to run the machine, for an hour or so, at full speed, without any current generated, so that the bearings can be worked smooth, and tested before the machine is put to actual work, and in case the bearings become heated, the defect must be remedied.

When a shunt-wound machine is to be connected in circuit, in multiple with another, it should be brought up to speed, and the field resistance gradually thrown out, until the pilot lamp, at the head of the machine, is practically at its full normal candle-power. The machine can then be connected in the circuit, and by watching the pressure indicator, and throwing out the field resistance, the machine will gradually perform its share of the work. When two or more shunt-wound machines are in multiple, all field and line connections should be traced and examined before starting.

102. In a compound-wound machine, the magnets are wound similarly to those of the ordinary shunt-wound machines, but have, in addition, extra coils in series, and arranged so that according to the number of lamps in use, each receives a certain amount of the current. A change in the number of lamps, connected in the circuit, will cause an increase of current in one, and a proportionate decrease in the other, so that the pressure is kept at a constant point, the series coils acting as an automatic regulator.

Any number of machines can be run in multiple, provided the pressure or electromotive force of all corresponds. Should the pressure on one be less than that of the others, it is liable to be run by them as a motor. With dynamos connected in multiple, the pressure of all is equal to that of any one machine, but the current, in amperes, is increased according to the capacity of all, for example: Three machines having a capacity of 100, 50 and 25 amperes, respectively, the pressure of each equals 125 volts, the combined output would equal 175 amperes at a pressure of 125 volts.

103. In stopping the machine, if running singly, slow the speed of the engine, which reduces the pressure in the dynamo ; throw the resistance coils in the field circuit, if hand regulator is used ; and just before the engine is stopped, break the line connection, and lift the brushes.

If the machine is connected in multiple with another, in order to stop or disconnect it from the circuit, regulate its amount of work so that it will be as small as possible, then break the line connection ; but the field must correspond in strength to that in the other machines, so that it may not be run as a motor.

This same method applies to machines connected to three-wire circuits, and also to compound-wound machines, when the equalizer is kept closed.

104. Two dynamos connected to the three-wire system, are practically similar to two dynamos connected in multiple ; the positive of one, and the negative of the other machine, are connected to the middle or neutral wire of the circuit. The other pole of each machine forms the positive and negative pole, respectively.

In starting the machines, each should be started separately, and not at the same time. When one machine is connected and running, the second machine, when up to the required pressure, can be connected in the circuit. The method of stopping, as stated, is the same as if in ordinary multiple.

105. In connecting machines in series, the positive pole of one connects with the negative pole of the next machine ; the current in amperes remains constant, but the pressure in volts increases with every additional machine ; the voltage of the machines need not be the same, but the current capacity of each must correspond with the others.

In constant current machines, when starting, it is only necessary to examine the connections, and have the circuit completed. The line or circuit should never be broken while running, as the field may "burn out," or if broken at the brushes, an arc will be created, that will burn the commutator. In stopping simply slow the speed, if possible, until the armature stops revolving.

If connected to shafting, or arranged so that speed cannot be stopped, then open the field circuit, but under no consideration, must the line be broken.

The method of starting and stopping is the same for a single series machine, or a number of them.

106. In setting up motors, the same care and attention must be given to all the parts and bear-

ing, as to dynamos. A blue print or diagram, showing the method of connecting the motor and the starting or regulating box, and directions for running same, should be provided.

In a series-wound motor, which is connected in an "arc light" circuit, the connections are simple. The motor is "cut-in" the circuit in the same manner as an arc lamp, a switch is provided and connected in the line (arc light hand switch), which is used for starting or stopping the motor; as in a series dynamo, the line must never be broken, or, the brushes must never be raised from the commutator.

Constant potential motors are wound in a manner similar to the ordinary shunt-wound dynamo, and with each motor is provided a starting box, composed of a number of coils, and constructed somewhat similar to a resistance box, used for regulating the field of shunt-wound dynamos. The wire forming one side of the circuit is connected to the post of the cross-bar, on the face-board of the starting box, and the cross-bar through a coil in the box forms the connection between one of the field wires and the line. The cross-bar also forms, through a number of coils, the connection between the line and one of the brush wires. The line should always be provided with a double-pole cut-out and switch. To start the motor, close the switch in the line, and turn the connecting strip or cross-bar on the starting box, until it rests on the first connecting plate on each side. Allow it to rest on these for an instant, so that the fields will be charged. As the armature begins to turn, move the cross-bar from plate to plate, which throws out the resistance, and the motor will then have attained its full speed. The movable bar on the starting box must not be allowed to rest on any of the connecting plates (with the exception of the first, for only a moment), but must be steadily moved to the last plate. This box is not to be used as a regulator, but is only for use when starting or stopping, as the capacity of the wires is not sufficient to carry the current for even a short space of time.

107. A speed regulator is constructed in a somewhat similar manner to the starting box, and the cross-section of the wires is such, that they can carry the current without unduly heating. In stopping, turn the handle, at the starting box, in the opposite direction to that when starting, and when it is brought to the last connection, the circuit is broken, although it is more preferable to allow the bar to rest on the connecting plate, next to the last, and break the connection at the line switch.

CHAPTER XXIV.

Pulleys.

108. In all machines, whether a dynamo or motor, the pulley is furnished with the machine. The width of face and diameter, have been determined by the maker. The size of the pulley on the dynamo has been determined, according to the speed and power required to drive it when at its full rated capacity. The pulley on the motor is determined by the rated speed at which the motor is to run, and its rated horse-power.

The pulleys on the machines should not be altered, or pulleys of other dimensions used ; and where conditions exist, which necessitate a change in size, the maker should be notified. He will either make the change, or forward instructions regarding the matter.

In dynamos driven by an engine, the dimensions of the driving wheel is usually determined by the electric lighting company or the maker of the engine. The diameter is determined by the number of revolutions per minute of the engine shaft, the revolutions, per minute, necessary for the capacity of the dynamo, and the diameter of the dynamo pulley. When the dynamo is belted to a pulley on a countershaft, the shaft must be considered as the shaft on the engine.

The face or width of a pulley is a trifle larger than the width of the belt.

109. To ascertain the required diameter of the driving pulley: Multiply the diameter of the dynamo pulley by the number of revolutions per minute required, and divide the product, by the number of revolutions per minute, of the shaft,

$$\frac{\text{Dia. of Dynamo Pulley} \times \text{required speed}}{\text{Revolutions of shaft, per minute,}} =$$

diameter of driving pulley.

Example: To determine the diameter of pulley to drive a dynamo, having a pulley 10 inches in diameter, and requiring 1,500 revolutions per minute. The revolutions of the shaft per minute = 225, therefore,

$$\frac{10 \times 1500}{225} = 66\tfrac{2}{3} \text{ inches} = \text{diameter of driving}$$

pulley.

In motors, the dimensions of the pulley being already determined by the maker, the speed at which the machinery or shaft is driven by the motor, depends on the speed of the motor, and the diameter of the pulleys. The diameter of the pulley on the machine or shaft to be driven at a certain speed, depends upon the speed and diameter of the motor pulley.

To ascertain the diameter of the driven pulley for any desired speed or number of revolutions per minute, multiply the revolutions per minute, of the motor pulley, by its diameter, and divide the product, by the number of revolutions, desired for the driven pulley.   Thus,

$$\frac{\text{Speed of motor} \times \text{diameter of motor pulley}}{\text{Desired number of revolutions for driven pulley}} =$$

Diameter, in inches, of driven pulley.

Example :   To determine the diameter of a pulley requiring 200 revolutions per minute, the rated speed of the motor being 1,200 revolutions per minute, and the diameter of pulley on same equals 9 inches, therefore,

$$\frac{1200 \times 9}{200} = 54 \text{ inches} = \text{diameter of driven}$$

pulley.

By these methods, the driving or driven pulley can be ascertained.

CHAPTER XXV.

Belting.

110. All belts and lacing should be the best procurable. The belt should be placed on the pulley, with the smooth side in contact. If the belt tends to run to either edge of the pulley, move the dynamo or motor in that direction, until the belt remains in the centre.

Do not skimp on the width or length of the belt, it being desirable to have it a trifle larger than is actually necessary rather than too small.

In dynamos or motors, the width of the belt is governed by the width of the face of the pulley on same. The pulley is usually one inch wider than the belt.

Belts driven horizontally give better satisfaction than those driven in a vertical position, as the arc of contact is increased in horizontal driving.

111. Avoid excessive strain, and protect the belts from dirt, exposure, extreme dampness or extreme dryness. Applying neats-foot oil occasionally will keep the belt soft and pliable.

112. A single belt traveling 1,000 feet per minute transmits one horse-power, provided the arc of contact equals 180°, or if it binds or touches one-half the surface of the circumference of the pulley.

Rule for ascertaining the width of belt for any desired horse-power:

$$\frac{\text{Width of belt} \times \text{speed of belt in ft.} \times \text{arc of contact}}{1000.}$$

= h. p.

CHAPTER XXVI.

ENGINES.

113. Although wiremen are not expected to be steam engineers, yet there are times when a knowledge of starting and stopping an engine may be very serviceable, as in the case of sudden sickness of the engineer in charge of plant, and where illumination is imperatively required.

The engines mostly used for driving dynamos are of the horizontal, high-speed type. The steam pressure is usually 80 pounds to the square inch, to obtain the required speed of revolution.

114. Before starting an engine, supply all the oil cups and "self"-lubricators with a sufficient amount of oil, and see that the "feed" is properly adjusted. Open the rear ports, or drips, usually located in the back of and under the cylinder, to allow any condensed water to run out. Set the engine on its "centre" and then slightly open the steam valve, to heat the piston and cylinder. When one side is "warmed up" turn the pulley over to the opposite "centre" and heat the opposite part of the cylinder, piston, etc. When the engine is heated, and

the steam gauge indicates the pressure required, open fully the exhaust valve, close the rear ports, or drip cocks, and slightly open the steam valve. Turn the driving pulley in the proper direction, and when the engine is running slowly, note whether it is running smoothly. Keep opening the steam valve gradually, until it is fully opened, and the engine is running at full speed. Attention can then be given to the starting of the dynamo.

115. While the engine is running, observe from time to time the amount of pressure indicated at the gauge, and watch the lubricators, etc.

If any loud noise or pounding be heard in the engine, shut down at once, as water may from some reason have been carried into the cylinder, and if the engine were kept running, the result would be that the cylinder-head would be blown out, and the engine be ruined.

To stop the engine, gradually close the steam valve, be careful that the engine is not stopped too sudden. Slowly bring the driving wheel to a stand-still. When the engine is stopped, close the feed of the lubricators, clean all parts of the engine, and cover it with the cloth provided for the purpose.

116. Under no consideration must it be attempted to repair any part or parts of the engine, and it is

suggested that, unless the engine is in good order, and the case urgent, the wireman should not in any manner handle the engine, unless he is an experienced engineer.

CHAPTER XXVII.

Conclusion.

117. Wiremen should familiarize themselves with the different kinds and qualities of material, so that they may be enabled to.get the best of any particular kind. In the matter of appliances, they should know just what is the best for the special purpose. New appliances, etc., are being placed on the market continually, and unless a knowledge of the same is acquired, antiquated, obsolete and inferior materials and devices will be used to the disadvantage of the customer. The writer would advise all wiremen to subscribe to some journal devoted exclusively to the discussion and publication of electrical engineering work. Hardly a week goes by but what some one or more articles treat on new methods of wiring and kindred work. It is very instructive and at the same time interesting. We are also enabled, through the same source to obtain information on the various appliances, and where the same can be purchased. We are better able to keep in line with all improvements in all branches ; as an instructor, the press has no equal, and it has

the advantage, moreover, of continually devising
new ways by which the wireman may get greater
profit from his experience and rise to larger
responsibilities and opportunities.

TABLES OF DIFFERENT GAUGES, WITH THEIR RESPECTIVE
DIAMETERS AND AREAS.

BROWNE & SHARPE.			BIRMINGHAM.		
No. of Gauge.	Diameter in Mils.	Area in C M = d².	No. of Gauge.	Diameter in Mils.	Area in C M = d².
4–0	.4600	211600	4–0	.454	206116
			3–0	.425	180625
3–0	.4096	167805			
2–0	.3648	133079	2–0	.380	144400
			0	.340	115600
0	.3249	105592			
			1	.300	90000
1	.2893	83694	2	.284	80656
2	.2576	66373	3	.259	67081
3	.2294	52634	4	.238	56644
			5	.220	48400
4	.2043	41742	6	.203	41209
5	.1819	33102	7	.180	32400
6	.162	26244	8	.165	57225
7	.1443	20822	9	.148	21904
8	.1285	16512	10	.134	17956
9	.1184	13110	11	.120	14400
10	.1019	10381	12	.109	11881
11	.0907	8226	13	.095	9025
12	.0808	6528	14	.083	6889
13	.072	5184	15	.072	5184
14	.0641	4110	16	.065	4225
15	.0571	3260	17	.058	3364
16	.0508	2581	18	.049	2401
17	.0452	2044	19	.042	1764
18	.0403	1624			
19	.0359	1253	20	.035	1225
20	.032	1024	21	.032	1024
21	.0285	820	22	.028	784
22	.0253	626	23	.025	625
23	.0226	510	24	.022	484
24	.0201	404	25	.020	400
25	.0179	320	26	.018	324

WEIGHT OF COPPER WIRE.

No.	One Thousand Feet.		One Mile.	
	B. & S.	B. W. G.	B. & S.	B. W. G.
0000	639.33	622.36	3375	3286
000	507.01	546.22	2677	2884
00	402.09	436.56	2123	2305
0	319.04	349.63	1684	1846
1	252.88	272.17	1335	1437
2	200.54	243.76	1058	1287
3	159.03	202.84	839	1071
4	126.12	171.21	665	904
5	100.01	146.40	528	773
6	79.32	124.43	418	657
7	62.90	97.92	332	547
8	49.88	82.39	263	435
9	39.56	66.29	209	350
10	31.37	54.36	166	287
11	24.88	43.56	131	230
12	19.73	35.98	104	190
13	15.65	27.27	83	144
14	12.41	20.83	65	110
15	9.84	15.72	52	83
16	7.81	12.88	41	68
17	6.19	10.18	33	$53\frac{3}{4}$
18	4.92	7.20	26	38
19	3.93	5.30	$20\frac{3}{4}$	28
20	3.09	3.60	$16\frac{1}{4}$	$19\frac{1}{2}$
21	2.45	3.09	13	$16\frac{1}{4}$
22	1.94	2.37	$10\frac{1}{4}$	$12\frac{1}{2}$
23	1.54	1.94	$8\frac{1}{8}$	$10\frac{1}{4}$
24	1.22	1.47	$6\frac{1}{2}$	$7\frac{3}{4}$
25	.97	1.22	$5\frac{1}{8}$	$6\frac{1}{2}$
26	.77	.95	4	5
27	.61	.75	$3\frac{1}{4}$	4
28	.48	$.61\frac{1}{2}$	$2\frac{1}{2}$	$3\frac{1}{4}$
29	.38	.50	2	$2\frac{5}{8}$
30	.30	.42	$1\frac{5}{8}$	$1\frac{1}{4}$

Table Showing Fractions of an Inch Reduced to Decimal Equivalents.

$\frac{1}{64}$	equals	.015625
$\frac{1}{32}$	"	.031250
$\frac{3}{64}$	"	.046875
$\frac{1}{16}$	"	.062500
$\frac{5}{64}$	"	.078125
$\frac{3}{32}$	"	.093750
$\frac{7}{64}$	"	.109375
$\frac{1}{8}$	"	.125000
$\frac{9}{64}$	"	.140625
$\frac{5}{32}$	"	.156250
$\frac{11}{64}$	"	.171875
$\frac{3}{16}$	"	.187500
$\frac{13}{64}$	"	.203125
$\frac{7}{32}$	"	.218750
$\frac{15}{64}$	"	.234375
$\frac{1}{4}$	"	.250000
$\frac{17}{64}$	"	.265625
$\frac{9}{32}$	"	.281250
$\frac{19}{64}$	"	.296875
$\frac{5}{16}$	"	.312500
$\frac{21}{64}$	"	.328125
$\frac{11}{32}$	"	.343750
$\frac{23}{64}$	"	.359375
$\frac{3}{8}$	"	.375000
$\frac{25}{64}$	"	.390625
$\frac{13}{32}$	"	.406250
$\frac{27}{64}$	"	.421875
$\frac{7}{16}$	"	.437500
$\frac{29}{64}$	"	.453125
$\frac{15}{32}$	"	.468750
$\frac{31}{64}$	"	.484375
$\frac{1}{2}$	"	.500000

DIMENSIONS, WEIGHT AND RESISTANCE OF BARE COPPER WIRE—AM. GAUGE.

Gauge No.	Diameter in Mils.	Sect. area in Circular Mils.	Weight. Lbs. per Foot.	Weight. Lbs. per Ohm.	Length—Feet. Per Lb.	Length—Feet. Per Ohm.	Resistance—Ohms. Per Foot.	Resistance—Ohms. Per Lb.	Gauge No.
0000	.46	211600.	.640525	13129.29	1.56122	20497.7	.000048786	.0000761656	0000
000	.40964	167805.	.507955	8256.95	1.9687	16255.27	.000061519	.00012111	000
00	.3648	133079.	.40284	5195.13	2.4824	12891.87	.0000775713	.000192559	00
0	.32486	105534.	.319457	3265.54	3.1808	10228.08	.000097818	.0008062	0
1	.2893	83694.	.253346	2054.015	3.94714	8107.49	.000123342	.000476956	1
2	.25763	66373.	.200015	1291.80	4.97722	6429.58	.0001555	.000774113	2
3	.22942	52634.	.158635	812.709	6.2765	5098.61	.000196132	.00123102	3
4	.20431	41743.	.125387	522.839	7.9141	4043.6	.000247804	.00191263	4
5	.18194	33102.	.10022	331.309	9.97983	3206.61	.000311856	.00311237	5
6	.16202	26251.	.0794616	202.063	12.5847	2342.89	.000393305	.00494896	6
7	.14428	20817.	.0630134	127.07	15.8696	2015.51	.000495905	.00785156	7
8	.12849	16510.	.0497757	79.9258	20.0097	1598.3	.000625975	.0125116	8
9	.11443	13094.	.039637	50.2966	25.229	1268.44	.00078897	.0198852	9
10	.10189	10382.	.0314256	31.6036	31.8212	1055.66	.000994487	.031642	10
11	.090742	8234.	.024925	19.882	40.1202	797.649	.0013337	.0502967	11
12	.080808	6530.	.0197665	12.5034	50.5906	632.555	.0015809	.0799753	12
13	.071961	5178.	.0156753	7.86819	63.7948	501.63	.0019985	.127173	13
14	.064084	4107.	.0124314	4.51033	80.4415	397.622	.0025187	.221713	14
15	.057068	3257.	.0098584	3.11015	101.4365	315.483	.00316975	.351528	15
16	.05082	2583.	.0078179	1.95501	127.12	250.184	.00399707	.511504	16
17	.045257	2048.	.0062	1.23013	161.29	198.409	.0050401	.812918	17
18	.040303	1624.	.004917	.773677	203.374	157.35	.0063533	1.29253	18
19	.08589	1288.	.0038991	.486524	255.468	124.777	.00801426	2.0554	19
20	.031961	1021.	.0030922	.305979	323.399	98.9538	.0101058	3.2692	20
21	.028462	810.	.0024522	.192439	407.815	78.479	.0127482	5.19671	21
22	.025847	642.	.0019448	.121087	514.198	63.236	.0160876	8.26197	22
23	.022571	509.	.0015431	.076105	648.452	49.8504	.0209883	13.13974	23
24	.0201	404.	.001223	.0478624	817.688	39.1365	.0255516	20.89828	24
25	.0179	320.	.0009699	.0301039	1031.086	31.0391	.0329184	33.2184	25
26	.01594	254.	.0007691	.0183719	1300.180	24.6131	.0406298	53.8247	26
27	.014195	201.	.0006099	.0119066	1639.49	19.5191	.0512946	85.994	27
28	.012641	159.8	.0004887	.0074748	2067.364	15.4798	.0646028	133.5368	28
29	.011357	126.7	.0003836	.0047087	2606.959	13.2654	.081464	212.373	29
30	.010025	100.5	.0003042	.00296174	3287.064	9.7855	.102717	337.639	30
31	.008928	79.7	.0002413	.0018306	4414.49	7.73143	.12951	536.7515	31
32	.00795	63.	.0001913	.00117183	5226.915	6.12348	.163884	853.788	32
33	.00708	50.1	.0001517	.00073676	6590.41	4.85575	.206943	1357.341	33
34	.006304	39.74	.0001203	.0004631	8312.8	3.84966	.25076	2159.361	34
35	.005614	31.5	.0000954	.000291272	10481.77	3.05806	.327541	3433.21	35
36	.005	25.	.00007569	.000183269	13314.16	2.4317	.41293	5456.45	36
37	.004453	19.8	.00006003	.000115398	16659.97	1.90586	.520601	8673.2	37
38	.003965	15.72	.00004759	.0000724741	21013.25	1.52992	.656635	13793.04	38
39	.003531	12.47	.00003774	.0000455829	26490.237	1.20777	.82797	21933.11	39
40	.003144	9.88	.00002992	.0000369903	33420.83	0.97984	1.04435	27041.4	40

www.ingramcontent.com/pod-product-compliance
Lightning Source LLC
Chambersburg PA
CBHW021808190326
41518CB00007B/504